# INSECTORAMA

### insect-

From the Latin *insectum*,
meaning *in several parts*.

### -orama

From the Greek *hórama*,
meaning *vision*.

**INSECTORAMA**
**The Marvelous World of Insects**

Text, illustrations, and layout:
Lisa Voisard · www.lisavoisard.ch

American adaptation: Dr. Mathilde Gaudreau, entomologist, Université de Montréal

Translation from French: Jeffrey K. Butt

Editor: Angela Wade

Proofreader: Sonia Curtis

ISBN: 978-3-0396-4016-4
First edition: March 2024
Printed in the UK

MIX
Paper | Supporting responsible forestry
FSC® C007785

© 2024 HELVETIQ (Helvetiq SA)
Avenue des Acacias 7
CH-1006 Lausanne, Switzerland
All rights reserved.

www.helvetiq.com

HELVETIQ Publishing has received a structural grant from the Swiss Federal Office of Culture for the years 2021–2025.

HELVETIQ
helvetiq.com

Front cover: small milkweed bug
Back cover: cabbage white, black swallowtail caterpillar, house fly, seven-spotted ladybug, yellow bumble bee, Carolina grasshopper

Painted lady

Lisa Voisard

# INSECTORAMA

## The Marvelous World of Insects

Translated by Jeffrey K. Butt

# Table of contents

Green darner

| | |
|---|---|
| What are insects? | 6 |
| Insect morphology | 8 |
| Many different orders | 10 |
| Features galore | 14 |

## 1 Portraits

| | |
|---|---|
| Towns, cities, and gardens | 19 |
| Countryside | 69 |
| Wetlands | 119 |
| Forests | 143 |
| Other astonishing insects | 168 |

## 2 Go on, take a closer look!

| | |
|---|---|
| Tips for getting started | 182 |
| How to identify an insect | 183 |
| Get the right gear | 184 |
| Rolling with the seasons | 186 |
| Towns, cities, and gardens | 188 |
| Countryside | 189 |
| Wetlands | 190 |
| Forests | 191 |
| Who goes there? | 192 |
| Why are some people afraid of insects? | 194 |

# 3 The life of insects

| | |
|---|---|
| Magical metamorphoses | 198 |
| Reproduction | 200 |
| Egg laying | 201 |
| Camouflage and mimicry | 203 |
| Communication and seduction | 204 |
| Defense | 207 |
| Predators | 208 |
| Miraculous migrators | 210 |
| Why are insects disappearing? | 214 |
| What insects do for the planet | 216 |
| What insects do for us | 217 |
| Let's do our part | 218 |

Snowberry clearwing

# What are insects?

## It's all about the legs

Insects are arthropods, a large group of creatures that also includes spiders, woodlice, millipedes, centipedes, scorpions, and crustaceans. But only those little critters with **six legs** can truly be called insects. They're also the only invertebrates capable of flight.

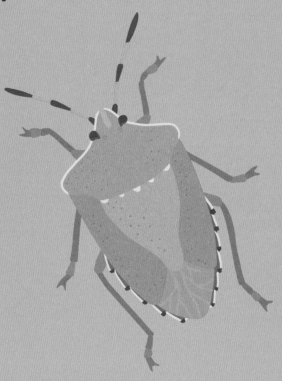

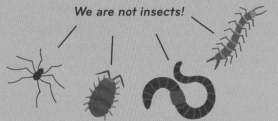

*We are not insects!*

## Striking transformations

In the animal kingdom, insects are the masters of metamorphosis. From an **egg** emerges a **larva**, which molts and transforms into a **nymph**, before finally becoming an **adult**.

## Small-statured

These tiny animals live their lives on an entirely different scale than we humans do. So, insect watching demands that you pay very close attention!

## Incredible diversity

Although very small and well-hidden, insects account for 85% of animal biodiversity. There are about 1.3 million known insect species alive today, and each year, about 10,000 new ones are discovered! As far as we know, beetles and butterflies are the most diverse types of insects.

## Ancient animals

Thanks to a strong aptitude for survival and a sophisticated social structure, as well as their resilience and ability to adapt quickly to environmental changes, insects have been around for a long, long, long time. The oldest insect fossils are around 300 million years old—which means they predate the dinosaurs!

# Insect morphology

An insect's body is separated into three parts: the head, the thorax, and the abdomen. All insects have six legs, two antennae, and many have one or two pairs of wings.

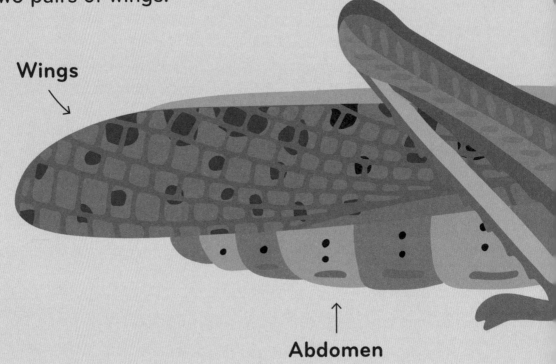

Wings

Abdomen

## Sight

Insects have well-developed eyesight. They generally have two compound eyes made up of "ommatidia"—small units that catch light and give them panoramic vision. Between these two eyes, there are often simple eyes, or "ocelli," which are smaller and help insects sense light intensity (as with dragonflies, bees, and grasshoppers, for example).

## Hearing

Insects' ears can be located in the most unexpected places. A cricket's ears are under its knees. Grasshoppers and cicadas have ears on their abdomens. A mosquito's ears are on its antennae, and the green lacewing has them at the base of its wings! Insects that cannot hear are alerted by vibrations.

Carolina grasshopper

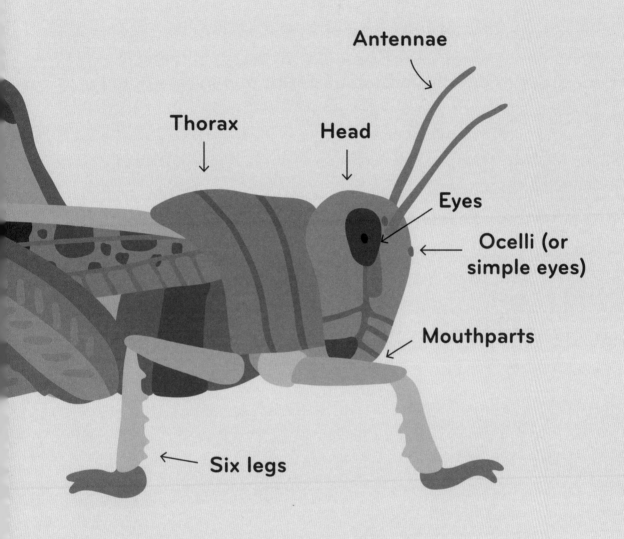

## Taste
Insects have mouthparts that allow them to taste their food. However, butterflies and flies can also taste with their feet.

## Smell
Insects' antennae often act as noses. Moths can detect scents from miles away, thanks to their feathery antennae.

## Touch
Insects are covered in tiny sensory hairs, which help them feel and analyze their surroundings. These hairs are found just about all over the body, but mainly on the legs, antennae, wings, and head.

# Many different orders

Ladybugs, dragonflies, and ants are all insects—yet, they look nothing alike! To understand insects better, we group them into orders based on their morphology, or physical form. There are nearly 30 orders, but these are the eight main ones:

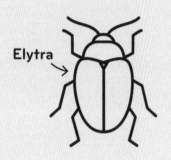

### Coleoptera
Ladybugs, beetles, long-horned beetles
**Around 400,000 species**

These insects possess two hardened wings called "elytra," which protect two other membranous, more delicate wings. The elytra give the insect stability in flight, while the membranous wings propel it forward. Species of this order often have powerful mouthparts, which they use to crush their food. These insects, generally called beetles, are known for their many different colors and patterns.

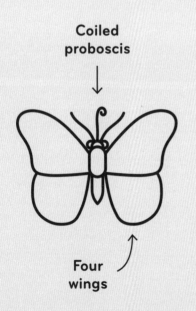

### Lepidoptera
Butterflies, moths
**Around 160,000 species**

These insects have four wings covered in tiny scales. There is no end to the variety of colors and patterns found in this order, and some butterfly wings are particularly stunning. Members of this order are pollinators that use a coiled "proboscis"—a trunk-like extension for feeding—to sip nectar from flowers.

## Diptera
Flies, mosquitoes, hover flies
Around 150,000 species

"Diptera" means two-winged. As these flies evolved, their second set of wings became much smaller and today their purpose is only to provide stability during flight. There is considerable diversity in this order in terms of the mouthparts used for feeding (for example, a proboscis, or a "stylet" used for piercing). These insects are also important pollinators and help recycle plant and animal waste.

## Hymenoptera
Bees, wasps, ants
Around 120,000 species

What distinguishes insects in this order from flies is their thin waist and four wings. Some of these pollinators are social insects that live in colonies.

Rostrum

Visible membranous wing tips

## Hemiptera
Aphids, bugs, cicadas

Around 100,000 species

These "true bugs" do not always look like they're related, but what they have in common is the structure of their mouthparts: all have a rostrum, or snout, for piercing and sucking. They are almost exclusively "phytophagous"—that is, they feed on plant material. Their first pair of wings is partially or completely membranous.

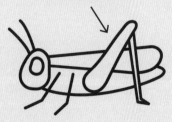

Hindlegs for leaping

## Orthoptera
Grasshoppers, crickets

Around 25,000 species

These insects are known for their ability to jump, which they do thanks to their powerful hindlegs. And they like to sing, too! Antenna size varies considerably among the species of this order—crickets have long ones, while grasshoppers have short ones.

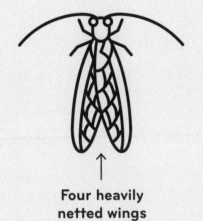

Four heavily netted wings

## Neuroptera

Lacewings, antlions, owlflies

Around 6,500 species

This order comprises insects with transparent wings that are heavily veined or netted, often forming a roof over their backs. Their larvae can be aquatic (water-based) or terrestrial (land-based), and are carnivorous thanks to mouthparts capable of crushing and grinding.

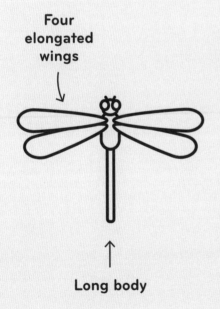

Four elongated wings

Long body

## Odonata

Dragonflies, damselflies

Around 6,000 species

These insects have long, slender bodies with four elongated wings. Their mouthparts allow them to crush food, and they have large, bulging eyes but tiny antennae. Their larvae are aquatic, but adults are terrestrial.

# Features galore

Insects can also be grouped based on a number of shared characteristics. Some feed on plants and others eat meat; some are active during the day, while others come out at night.

## Activity

Insects that are active during the day are "diurnal," whereas those that come out at night are "nocturnal." Others are "crepuscular"—they can be seen as the sun is rising or setting.

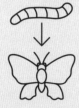

## Metamorphosis

The development of most insects plays out over several stages: from egg to larva to nymph to adult.

→ For more information, go to page 198

## Caution!

Some insects bite, sting, or pinch, but many are harmless. If you're unsure, it's best to leave them alone, as most insects only defend themselves when they feel threatened.

## Migration

Some insect species migrate before winter, heading off in search of warmer climes. They might be small, but they can travel vast distances!

→ For more information, go to page 210

# Five feeding adaptations

## Phytophagous
These insects are "herbivores," which means they eat living plants. Their diets may include leaves, grass, flowers, nectar, pollen, fruit, seeds, roots, sap, and wood. Nearly half of all insect species belong to this group.

## Saprophagous
These insects are "detritivores": they feed on decomposing plant and animal matter, as well as on fungi and excrement (poo).

## Xylophagous
This type of herbivore eats only wood (dead or alive). These insects break down old roots and house frames, and even attack trees by boring holes in the bark.

## Carnivorous
"Carnivores" feed on animal matter. They suck blood or grow inside other animals. They sometimes eat other insects, and some even practice cannibalism.

## Omnivorous
"Omnivores" will eat just about anything!

# Four habitats

Towns, cities, and gardens

Countryside

Wetlands

Forests

# 1

# Portraits

Some are stout, others are slender; some have powerful legs, others have long antennae; some feed on wood, others prefer leaves; some are solitary, others are social... Insects are extremely diverse. Read on to discover how different they look and learn about all the different ways they behave!

**Aerial yellowjacket**
p. 21

**German cockroach**
p. 25

**European earwig**
p. 29

**European honey bee**
p. 33

**Golden-eyed lacewing**
p. 39

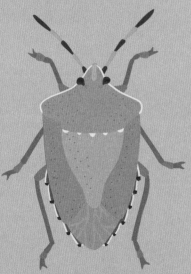

**Green stink bug**
p. 45

**House fly**
p. 49

**Larder beetle**
p. 53

# Towns, cities, and gardens

**Seven-spotted ladybug**
p. 61

**Margined calligrapher**
p. 57

**Pea aphid**
p. 65

# Aerial yellowjacket

A bad rap

**Order**
Hymenoptera

**Activity**
Diurnal

**Habitat**
Towns, cities, gardens, and countryside

With its yellow-striped suit and dreaded stinger, this species of wasp needs no introduction! In summer, the aerial yellowjacket is an unwelcome picnic guest, but if we are seeing it more often, that's because its natural food sources are disappearing due to climate change and dwindling biodiversity.

Only females are capable of stinging, but they do not die afterward, as bees do. Although considered a nuisance, wasps play an important role in the ecosystem: they pollinate flowers, help recycle decomposing plant matter, and keep other insect populations in check by hunting them to feed their larvae. They are also a source of food for birds.

Wasps shelter underground or build impressive nests out of chewed wood fiber.

**Length**
Female worker: 10 to 14 mm

Male: 13 to 17 mm

Queen: 16 to 19 mm

**Scientific name**
*Dolichovespula arenaria*

**Caution!**
May sting

## Metamorphosis

**Egg**
5 to 8 days

**Larva**
2 weeks

**Nymph**
1 to 3 weeks

**Adult**
Female worker/male:
10 to 30 days
Queen: 1 year

## Food

The larvae are **carnivorous**. Adults bring them small insects, spiders, and pieces of meat to eat.

Adults are **omnivorous**, feeding on nectar, fruit, sugary juices, and sometimes small insects.

## Not to be confused with...

The **European honey bee**, which has a brown coloration and a furry thorax.

## Geography
Found only in North America.

## Migration
The aerial yellowjacket is sedentary (it does not migrate).

# Observation guide

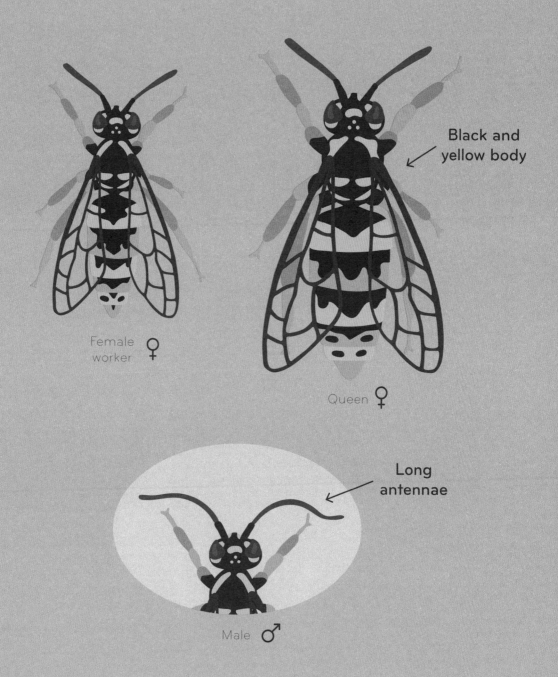

Female worker ♀

Black and yellow body

Queen ♀

Long antennae

Male ♂

## Where and when to see it
In parks and gardens where there is human activity, especially during the summer.

## Ease of observation

# German cockroach

Unloved and unwanted

**Order**
Blattodea

**Activity**
Nocturnal

**Habitat**
Towns and cities

No argument here—if there's one insect that puts people off, it's the cockroach. We associate it with filth, sewers, and garbage. But don't blame the cockroach if that's where it finds its food, which includes leftover bread, fatty foods, old cardboard, and even toothpaste! Cockroaches can be a sign of poor sanitation or too much dampness in a home. Good ventilation and keeping crumbs off the floor are ways to avoid infestations. If you are tasked with getting rid of a cockroach, be careful not to crush it, because it releases a foul odor that attracts more cockroaches. Better just to put it outside.

Cockroaches tend to come out at night and don't like the cold, which is why they are often found indoors. Cockroaches have flat bodies that help them slip into small cracks. They walk with pace and use their wings to make small, awkward hops to escape.

**Length**
~15 mm (excluding the antennae)

**Scientific name**
*Blattella germanica*

**Caution!**
May transmit bacteria or trigger allergies

## Metamorphosis

**Ootheca
(or egg case)**
3 to 4 weeks

**Larva**
1 to 3 weeks

**Adult**
3 to 7 months

## Food

The larvae and adults are **omnivorous**, feeding on starch, leftovers, fatty and sugary foods, and old cardboard.

## Not to be confused with...

The **American cockroach**, which is much larger (~35 mm) and has two dark, round spots on its thorax, as opposed to two stripes like the German cockroach.

## Geography
Found worldwide.

## Migration
The German cockroach is sedentary (it does not migrate).

# Observation guide

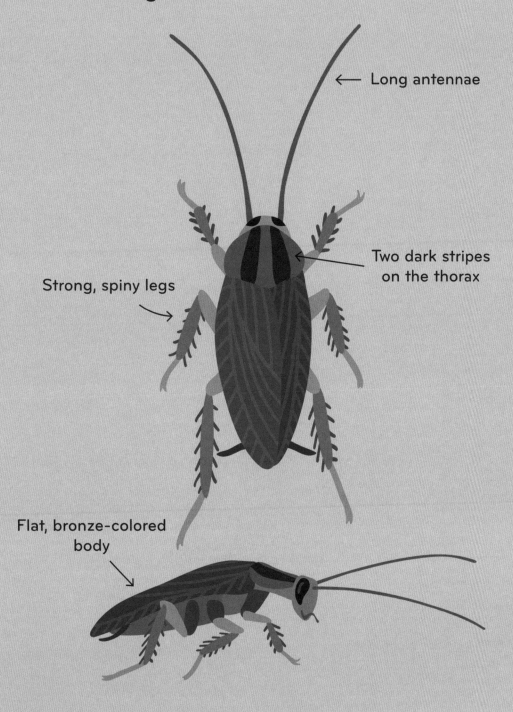

Long antennae

Two dark stripes on the thorax

Strong, spiny legs

Flat, bronze-colored body

### Where and when to see it
In damp recesses in the home, and in cellars and garages. Active at night, from spring to fall.

### Ease of observation

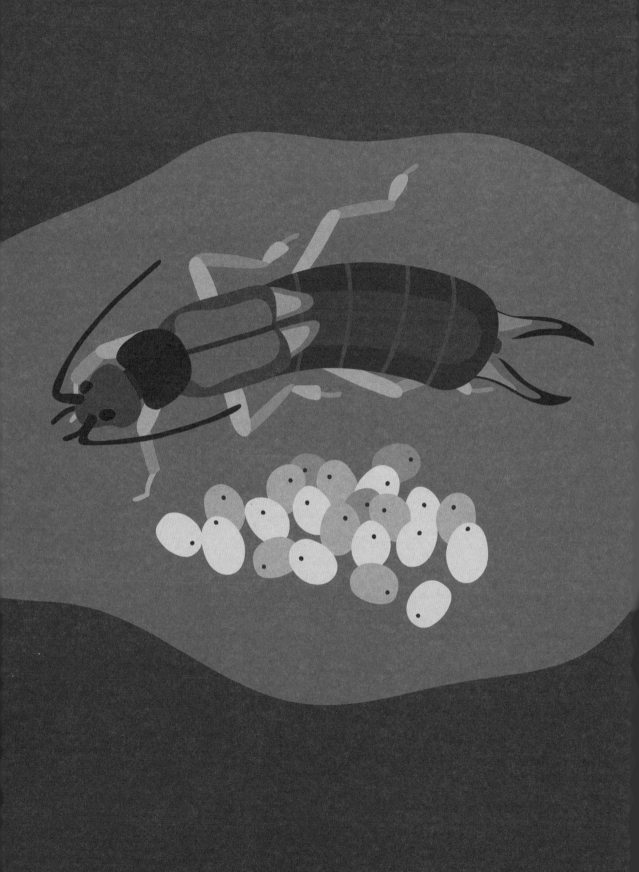

# European earwig

One busy mama

**Order**
Dermaptera

**Activity**
Nocturnal

**Habitat**
Towns, cities, gardens, and countryside

"Wig" comes from an Old English word meaning insect or beetle, and some people believe the creepy name of this insect has to do with it crawling into our ear canals. But rest assured—earwigs have no interest whatsoever in our ears! Scientists believe the name has more to do with the unique shape of the wing, which resembles an ear when unfolded. Adding to our fear of the European earwig are the tiny pincers ("cerci") at the end of its abdomen, but their purpose is to scare off predators. The insect hardly ever uses them to pinch—and when it does, it's painless!

The European earwig comes out at night. In the garden, this insect can be found hanging out under stones and flowerpots. It eats aphids and ripe fruit, and is especially fond of sweet flower petals.

Females spend a great deal of time caring for their eggs and larvae, until they become independent—a rare trait in the insect world!

**Length**
10 to 20 mm

**Scientific name**
*Forficula auricularia*

**Caution!**
May pinch (but only rarely and without causing pain)

## Metamorphosis

**Egg**
10 days
to 3 months

**Larva**
1 to 2 months

**Adult**
5 to 12 months

## Food

The larvae and adults are **omnivorous**, feeding on aphids, ripe or rotting fruit, and flower petals.

## Not to be confused with...

The **rove beetle**, which, depending on its color variation, can look like an earwig. This insect is a member of the Coleoptera order (that is, it's a beetle, as its name suggests).

## Geography
Native to Europe but also found in Australia and North America, where they are considered invasive.

## Migration
The European earwig is sedentary (it does not migrate).

# Observation guide

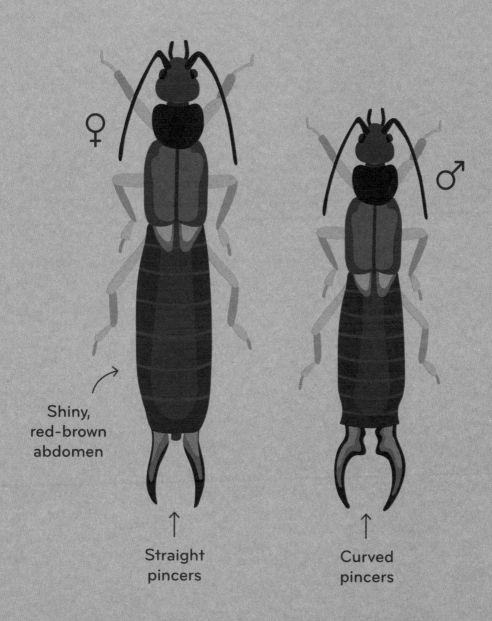

Shiny, red-brown abdomen

↑ Straight pincers

↑ Curved pincers

## Where and when to see it
In damp areas, under dead wood, stones, and flowerpots. Especially active during summer.

## Ease of observation

# European honey bee

Hive alive!

**Order**
Hymenoptera

**Activity**
Diurnal

**Habitat**
Towns, cities, gardens, and countryside

Beehives are abuzz with activity! The queen bee's job is to lay eggs, while female worker bees collect nectar, feed the larvae, and do the housekeeping. Male bees, called drones, are responsible for fertilizing the queen and keeping the eggs and larvae warm.

The primary purpose of honey-bee honey is to feed the next generation, and it uses beeswax to build the honeycombs that serve as both pantry and nursery. Honey bees also secrete a very nutritious substance, called royal jelly, that's reserved for the larvae and future queens. Because European honey bees have high yields, beekeepers have been managing their colonies for thousands of years.

Only females are capable of stinging. When they do, their stinger stays in the skin, which rips away a piece of their abdomen. This causes them to die soon after.

**Length**
Female worker:
10 to 13 mm

Male:
13 to 16 mm

Queen:
15 to 20 mm

**Scientific name**
*Apis mellifera*

**Caution!**
May sting

## Metamorphosis

**Egg**
3 days

**Larva**
10 days

**Nymph**
8 days

**Adult**
Female worker:
6 to 8 weeks
Male: 1 to 3 months
Queen: 4 to 7 years

## Food

The larvae and adults are **phytophagous**, feeding on nectar, pollen, and royal jelly.

## Not to be confused with...

The **aerial yellowjacket**, which has a slimmer waist and a hairless yellow and black body.

## Geography
Native to Europe, but now found worldwide (except polar regions).

## Migration
The European honey bee is sedentary (it does not migrate).

# Observation guide

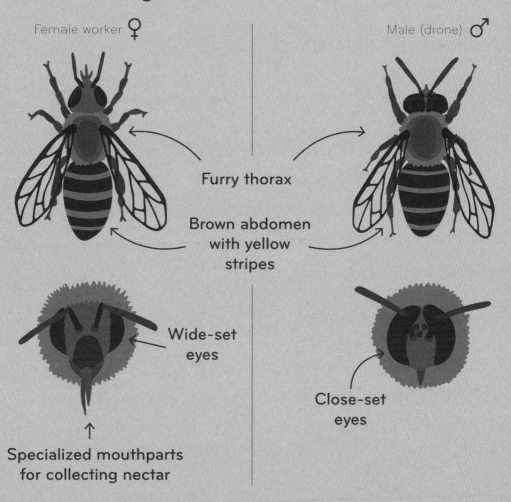

Female worker ♀ — Furry thorax, Brown abdomen with yellow stripes, Wide-set eyes, Specialized mouthparts for collecting nectar

Male (drone) ♂ — Close-set eyes

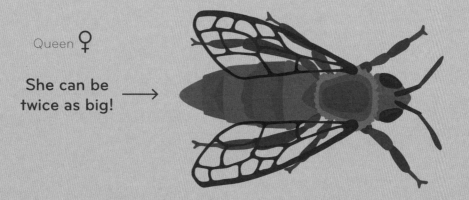

Queen ♀ — She can be twice as big!

### Where and when to see it
On flowers and in beehives, from early spring to the end of summer.

### Ease of observation

# A bee ballet

When a bee finds food, it can signal the location to its friends once back in the hive! How? With a dance!

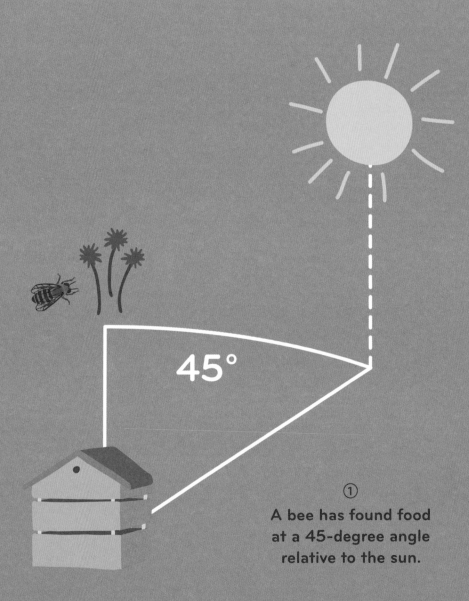

① A bee has found food at a 45-degree angle relative to the sun.

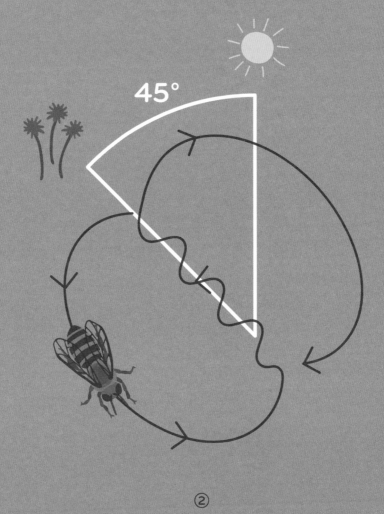

②
The bee reports the discovery by performing a figure-of-eight "waggle dance" on the beehive's honeycombs. The bee moves from side to side along the center line to indicate the direction of the food relative to the sun. All the other bees have to do is head in the right direction!

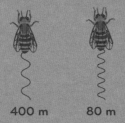

The bee can also indicate the distance to the food: the closer the food source, the faster the waggle.

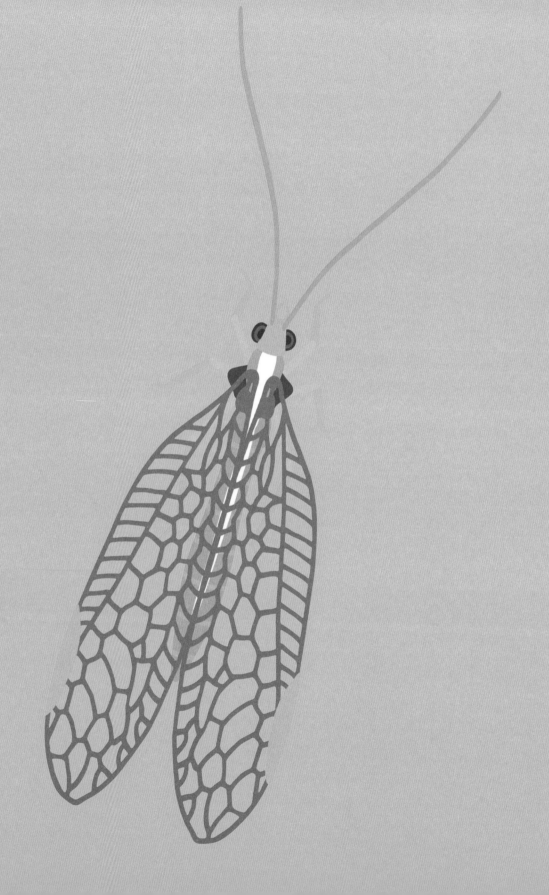

# Golden-eyed lacewing

## The delicate one

**Order**
Neuroptera

**Activity**
Crepuscular

**Habitat**
Towns, cities, gardens, and countryside

The golden-eyed lacewing is a small, delicate insect that is completely green. As a member of the Neuroptera order, its wings are heavily netted. Its name has to do with its shimmery eyes that change color when you look at them from different angles.

The larvae are carnivorous. With their long jaws, they catch aphids and mealy bugs—insects that are destructive to crops. In this way, golden-eyed lacewing larvae can be used as a natural alternative to garden insecticides.

Golden-eyed lacewings have a very interesting way of laying their eggs. After choosing the underside of a leaf where she wants to deposit her eggs, the female stretches her abdomen to produce a thin, yet sturdy, filament (thread). She then hangs her eggs from the end of these "pedicels," and the whole thing looks like half a cotton swab.

**Length**
10 to 15 mm (excluding the antennae)

**Wingspan**
23 to 30 mm

**Scientific name**
*Chrysopa oculata*

## Metamorphosis

**Egg**
3 to 6 days

**Larva**
15 to 20 days

**Nymph in its cocoon**
10 to 14 days

**Adult**
7 to 30 days

## Food

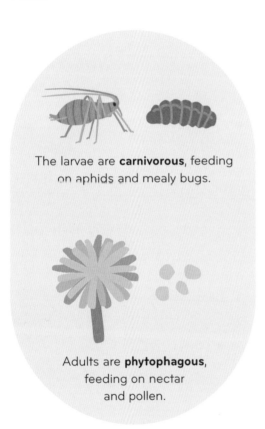

The larvae are **carnivorous**, feeding on aphids and mealy bugs.

Adults are **phytophagous**, feeding on nectar and pollen.

## Not to be confused with...

The **posterior brown lacewing**, which has a beige and brown body and wings covered with hair.

## Geography

Found in Canada, the United States, and Mexico.

## Migration

The golden-eyed lacewing is migratory.

# Observation guide

Entirely green body

Long antennae

Iridescent (shimmery, color-changing) eyes

Heavily netted wings

### Where and when to see it
On flowers and windowpanes, from spring to the end of summer.

### Ease of observation

# Hanging eggs

The golden-eyed lacewing lays strange-looking eggs on the underside of leaves. Why strange? Because they look like half a cotton swab!

# Green stink bug

A voracious appetite

**Order**
Hemiptera

**Activity**
Diurnal

**Habitat**
Towns, cities, gardens, countryside, and forests

The green stink bug looks like a small green shield and is known for the putrid odor it emits whenever it feels threatened.

This food-obsessed bug loves to feast on vegetable gardens and crop fields! It sucks the sap from the flower buds of tomato, eggplant, bell pepper, and soybean plants, and other summer fruits and vegetables.

When winter sets in, the green stink bug tucks itself away under tree bark to escape the cold.

During its metamorphosis, the green stink bug goes through several larval stages. The larvae are black and dotted with little splashes of color. This insect doesn't undergo full metamorphosis, as it has no nymphal stage.

**Length**
13 to 19 mm

**Scientific name**
*Chinavia hilaris*

## Metamorphosis

**Egg**
5 to 20 days

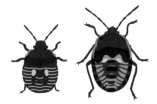
**Larval stages**
25 to 40 days

**Adult**
2 months

## Food

## Not to be confused with...

The **green burgundy stink bug**, which is a little smaller. Its body is brown and green, but its back lacks the green stink bug's yellow dots.

The larvae and adults are **phytophagous**, feeding on tomato, eggplant, bean, and soybean sap.

## Geography
Found in the United States, Canada, and Mexico.

## Migration
The green stink bug is sedentary (it does not migrate).

# Observation guide

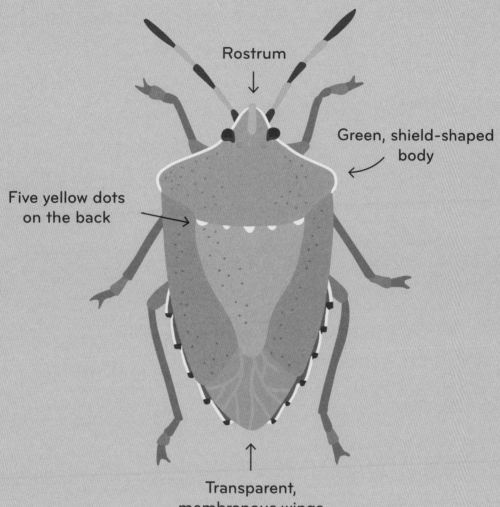

Rostrum

Green, shield-shaped body

Five yellow dots on the back

Transparent, membranous wings

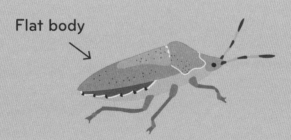

Flat body

**Where and when to see it**
On plants and tree bark during spring and summer.

**Ease of observation**

# House fly

## The clean freak

**Order**
Diptera

**Activity**
Diurnal

**Habitat**
Towns, cities, gardens, countryside, and forests

Nothing conjures up an icky feeling like the thought of a house fly, but it's actually an exceptionally clean insect! Its body is covered in tiny hairs that allow it to smell, taste, and feel vibrations. If a house fly becomes covered in dust, it loses all its bearings, which is why it's always rubbing its feet, body, and wings.

Thanks to their panoramic vision and ability to take off quickly, house flies are masters of escape—they're virtually impossible to catch. They also have little hooks on their feet that allow them to walk on the ceiling without falling. House flies cannot eat solid food, so they use their saliva to dissolve it, then lap up the liquid using their proboscises.

House-fly larvae are called maggots, and many people find them disgusting. However, they play a role in the natural recycling of organic matter.

**Length**
6 to 11 mm

**Wingspan**
~12 mm

**Scientific name**
*Musca domestica*

## Metamorphosis

**Egg**
1 to 2 days

**Larva**
5 to 6 days

**Nymph (or pupa)**
5 days

**Adult**
2 weeks to 2 months

## Food

The larvae are **saprophagous**, feeding on decomposing plant and animal matter, as well as fecal matter.

Adult house flies are **omnivorous**, feeding on nectar and pollen from flowers and just about anything sweet.

## Not to be confused with...

The **blue bottle fly**, which is stouter and has a blue body.

## Geography
Found worldwide.

## Migration
The house fly is sedentary (it does not migrate).

# Observation guide

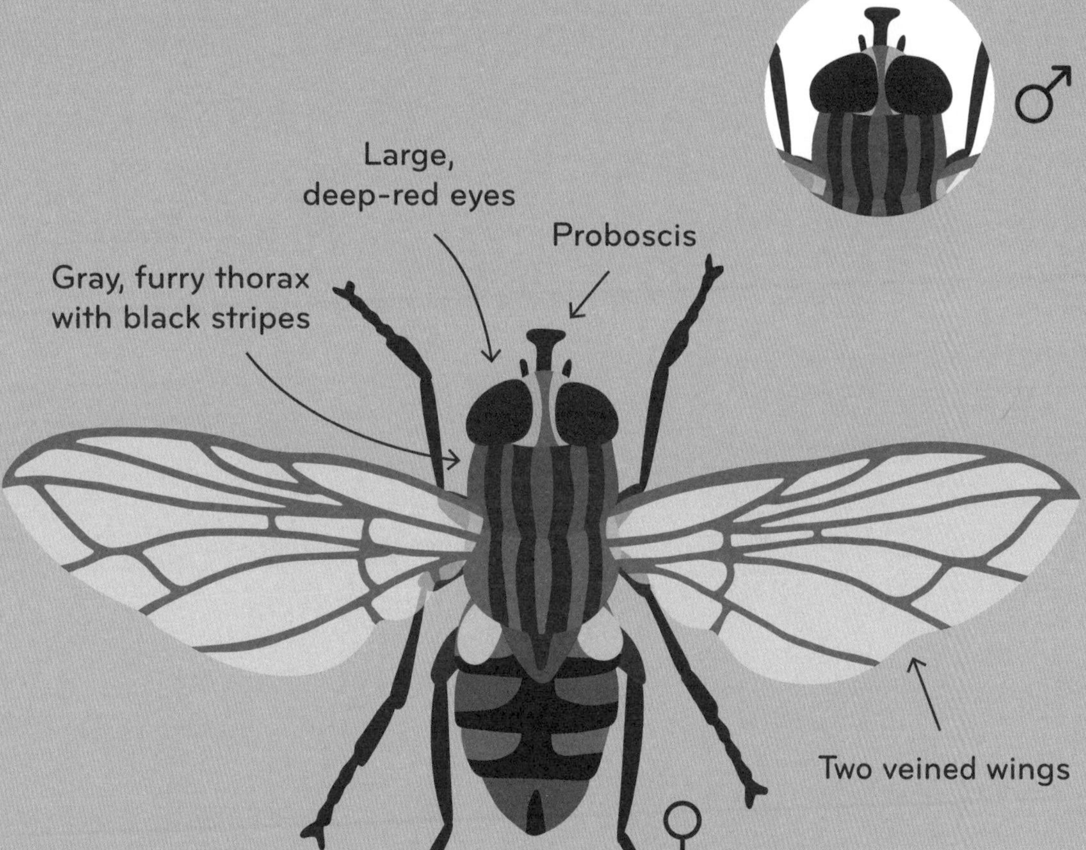

Males have close-set eyes ↓

♂

Large, deep-red eyes

Proboscis

Gray, furry thorax with black stripes

Two veined wings

♀

Mustard-yellow abdomen with black stripes

## Where and when to see it
Pretty much everywhere! House flies are at home in houses and on farms, and are active from spring to fall.

## Ease of observation
●●●●○

# Larder beetle

Museum friend and foe

**Order**
Coleoptera

**Activity**
Nocturnal

**Habitat**
Towns, cities, and forests

Larder beetle larvae can eat like horses! They feed on animal flesh and help rid attics and farms of small, dead animals—rodents, birds, spiders, and insects all make their day!

It's not uncommon for natural history museums to display animal skeletons to educate the public. Larder beetles are used to rid animal carcasses of their flesh to produce a perfectly clean, intact skeleton. However, they would cause some serious damage if they got inside the taxidermy display cases!

In the home, larder beetles are also known to destroy old books, rugs, and woodwork.

**Length**
5 to 8 mm

**Scientific name**
*Dermestes lardarius*

## Metamorphosis

**Egg**
1 to 2 weeks

**Larva**
2 to 12 weeks

**Nymph**
2 weeks

**Adult**
Up to 1.5 years

## Food

The larvae and adults are primarily **saprophagous**, feeding on the skin, feathers, and fur of animal carcasses, and occasionally on old books, fabrics, and seeds.

## Not to be confused with...

The **varied carpet beetle**, which is also found in houses but is rounder in shape and has black, orange, and white speckles.

## Geography
Found in the United States, Canada, Mexico, and Europe.

## Migration
The larder beetle is sedentary (it does not migrate).

# Observation guide

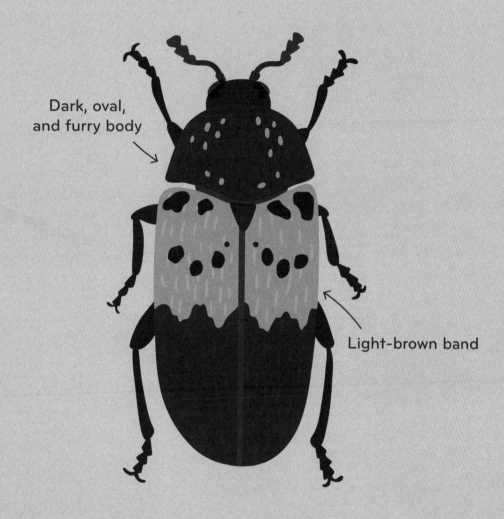

Dark, oval, and furry body

Light-brown band

**Where and when to see it**
In carpets and pantries, as well as in trees, year-round.

**Ease of observation**

# Margined calligrapher

The hoverer

**Order**
Diptera

**Activity**
Diurnal

**Habitat**
Towns, cities, gardens, countryside, and forests

It might look like a little wasp, but the margined calligrapher doesn't sting. It belongs to the Diptera order, along with flies and mosquitoes, and is recognizable by its hovering, erratic flight. The margined calligrapher lays its eggs in aphid colonies so that the larvae, once hatched, have a ready-made meal.

Both adult and larval margined calligraphers spend the winter in jumbles of ivy. During that time, they don't need to eat because their bodies go into idle mode—a state of dormancy known as "diapause." As soon as the weather starts to warm up, they feed on hazel and willow tree pollen.

**Length**
5 to 7 mm

**Wingspan**
12 to 14 mm

**Scientific name**
*Toxomerus marginatus*

## Metamorphosis

**Egg**
2 to 7 days

**Larva**
15 to 20 days

**Nymph (or pupa)**
10 to 15 days

**Adult**
1 month

## Food

The larvae are **carnivorous**, feeding on aphids and caterpillars.

Adults are **phytophagous**, feeding on the nectar from flowers.

## Not to be confused with...

The **aerial yellowjacket**, which is larger and flies more smoothly than the calligrapher, which tends to hover.

## Geography
Found in Canada, the United States, Mexico, and Central America.

## Migration
The margined calligrapher is sedentary (it does not migrate).

# Observation guide

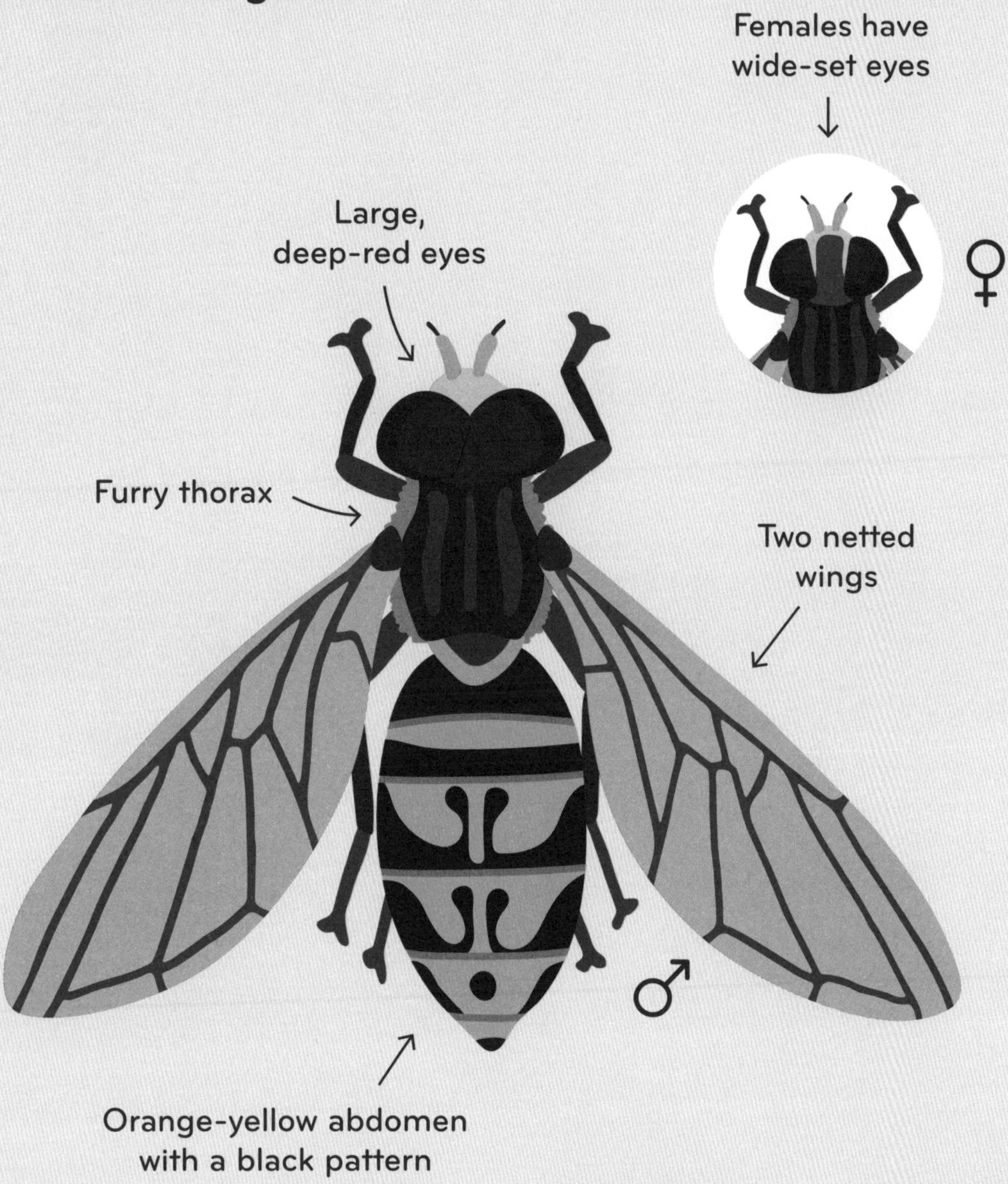

Females have wide-set eyes ↓

♀

Large, deep-red eyes ↙

Furry thorax →

Two netted wings ↙

♂

↑ Orange-yellow abdomen with a black pattern

### Where and when to see it
On flowers, stones, and dirt roads, and along forest edges. Active from spring to fall.

### Ease of observation
●●●○○

# Seven-spotted ladybug

The darling of the garden

**Order**
Coleoptera

**Activity**
Diurnal

**Habitat**
Towns, cities, gardens, and countryside

This little round beetle is one of the most popular insects, but underneath its cute exterior lurks a true predator. During the larval and adult stages, this ladybug loves to feast on aphids and other small insects—no wonder it's often used in gardens to help control pests.

Seven-spotted ladybugs undergo a fascinating metamorphosis: from egg to adult, they never look the same. The larvae are black and spiny—nothing like the adults—while nymphs, which are often seen on bush leaves, look like tiny brains.

According to folklore, ladybugs are good luck, so if one lands on you, just make a wish!

**Length**
5 to 8 mm

**Scientific name**
*Coccinella septempunctata*

## Metamorphosis

**Egg**
3 to 5 days

**Larva**
2 to 3 weeks

**Nymph**
8 days

**Young adult**
A few hours

**Adult**
1 to 3 years

## Food

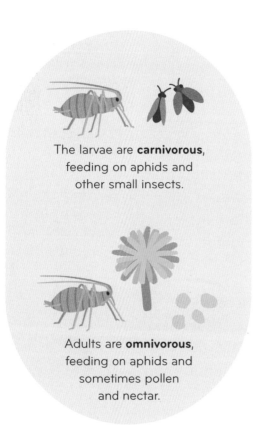

The larvae are **carnivorous**, feeding on aphids and other small insects.

Adults are **omnivorous**, feeding on aphids and sometimes pollen and nectar.

## Not to be confused with…

The **harlequin ladybug (or Halloween beetle)**, which has an M-shaped mark on its thorax. It has red or pale orange coloration and a variable number of spots.

## Geography
Native to Europe, but now found worldwide.

## Migration
The seven-spotted ladybug is sedentary (it does not migrate).

# Observation guide

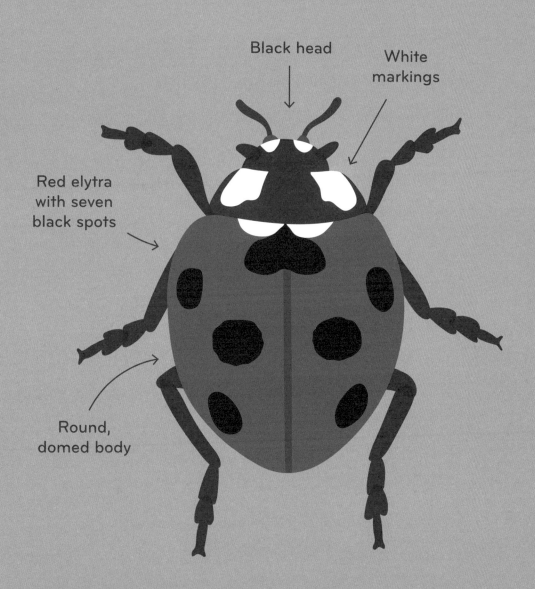

Black head

White markings

Red elytra with seven black spots

Round, domed body

### Where and when to see it
Larvae and nymphs can be found on leaves in spring. Adults can be seen throughout summer in grasses and bushes.

### Ease of observation

# Pea aphid

A miniature glutton

**Order**
Hemiptera

**Activity**
Diurnal

**Habitat**
Towns, cities, gardens, and countryside

This tiny insect feeds on many plants, including peas, clover, rosebushes, and lettuce. It just sticks its rostrum in and sucks out the sap. Doing so weakens the plants, which is a real nuisance for crop growers! Fortunately, aphids are on the menu for many other insects, so their numbers are kept in check.

Ants raise aphid colonies, exchanging good care for honeydew, which is sweet-tasting aphid excrement. In fact, a lot of insects are fans of honeydew, and bees even sometimes make honeydew honey.

At certain times of year, female aphids have the strange ability to reproduce all by themselves! This phenomenon, whereby a baby is created without an egg being fertilized, is known as "parthenogenesis."

**Length**
2.5 to 4.5 mm

**Scientific name**
*Acyrthosiphon pisum*

## Metamorphosis

**Egg**
3 months

**Larva**
8 days

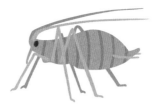
**Adult**
15 to 20 days

## Food

The larvae and adults are **phytophagous**, feeding on the sap of bean and pea plants.

## Not to be confused with...

The **rose aphid**, which has various black body parts.

## Geography

Native to Europe, but now found worldwide.

## Migration

The pea aphid is sedentary (it does not migrate).

# Observation guide

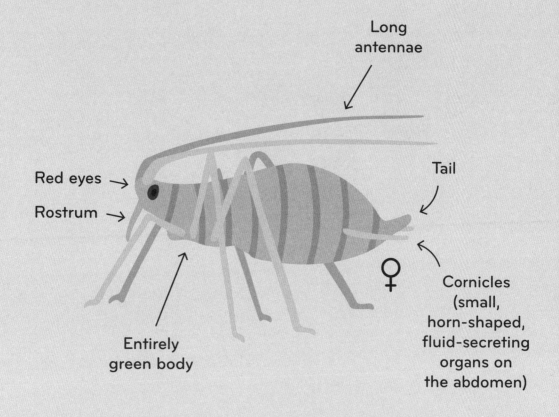

Long antennae

Tail

Cornicles (small, horn-shaped, fluid-secreting organs on the abdomen)

Red eyes

Rostrum

Entirely green body

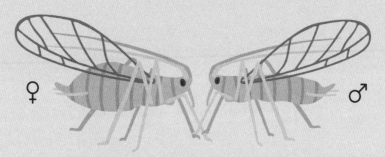

When reproduction involves a male and a female, both sexes sport wings.

## Where and when to see it
Under tree leaves, inside flowers, or on stalks, from spring to the end of summer.

## Ease of observation

**Fork-tailed bush katydid**
p. 93

**Monarch**
p. 97

**Small milkweed bug**
p. 107

# Countryside

**Painted lady**
p. 103

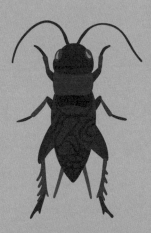

**Spring field cricket**
p. 111

**Yellow bumble bee**
p. 115

# Black swallowtail

A gifted glider

**Order**
Lepidoptera

**Activity**
Diurnal

**Habitat**
Countryside, towns, cities, and gardens

The black swallowtail is a large black butterfly adorned with prominent light-yellow and blue markings. The black swallowtail has two slender tails at the tips of its hindwings. It rarely has to flap its wings; instead, it takes advantage of the wind to glide about.

The caterpillar is green and plump, and it can often be seen on celery and carrot plants. It defends itself with its "osmeterium"—a fork-shaped organ located on its head that produces foul-smelling, bad-tasting secretions to keep predators, such as birds, ants, and even spiders, at bay.

When the caterpillar is ready to turn into a chrysalis, it finds a sturdy stalk and begins to spin a silken thread. The thread wraps around the caterpillar like a little belt and holds it in place.

**Length**
70 to 90 mm

**Wingspan**
70 to 100 mm

**Scientific name**
*Papilio polyxenes*

## Metamorphosis

**Egg**
1 week

**Caterpillar**
1 to 4 weeks

**Chrysalis**
2 weeks
to 8 months

**Adult**
1 to 3 weeks

## Food

The larvae and adults are **phytophagous**, feeding on carrot and celery plants. The larvae eat the leaves, while adults prefer the nectar.

## Not to be confused with...

The **Canadian tiger swallowtail**, whose yellowish wings have black streaks and longer tails.

## Geography
Found in Canada, the United States, Mexico, and northern areas of South America.

## Migration
The black swallowtail migrates occasionally.

# Observation guide

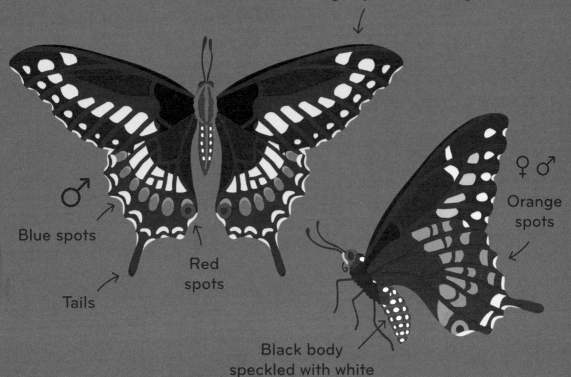

Black wings adorned with light-yellow markings

♂ Blue spots

Tails

Red spots

♀♂ Orange spots

Black body speckled with white

♀

The female has fewer light-yellow spots, but more blue

### Where and when to see it
In vegetables gardens, meadows, and fields, from the end of winter to fall.

### Ease of observation

# Cabbage white

Radiantly white

**Order**
Lepidoptera

**Activity**
Diurnal

**Habitat**
Countryside, towns, cities, gardens, and forests

Although native to Europe, the cabbage white is a very common butterfly that can be found pretty much everywhere, except South America. It has adapted to every environment: from flower gardens, meadows, and fields to city parks—anywhere there are beautiful flowers full of nectar. The cabbage white can even be found in the mountains, up to altitudes of 3,000 meters (9,800 feet).

In spring, adults begin to emerge from the cocoons that served as their winter homes. Couples form, and then the female finds a cabbage, mustard, or turnip leaf to lay her eggs on. When the caterpillar appears, it eats its own eggshell before feasting on the surrounding leaves.

The male cabbage white has one black spot on its forewings, while the female has two.

**Length**
20 to 30 mm

**Wingspan**
30 to 50 mm

**Scientific name**
*Pieris rapae*

## Metamorphosis

**Egg**
5 to 7 days

**Caterpillar**
2 to 3 weeks

**Chrysalis**
1 to 2 weeks

**Adult**
1 to 3 weeks

## Food

The larvae and adults are **phytophagous**. The larvae feed on cabbage, mustard, and turnip leaves, while adults feed on flower nectar.

## Not to be confused with...

The **common sulphur**, which has wings edged in black. Its color is often sallow, although the female is sometimes white.

## Geography
Found worldwide, except South America.

## Migration
The cabbage white is migratory.

# Observation guide

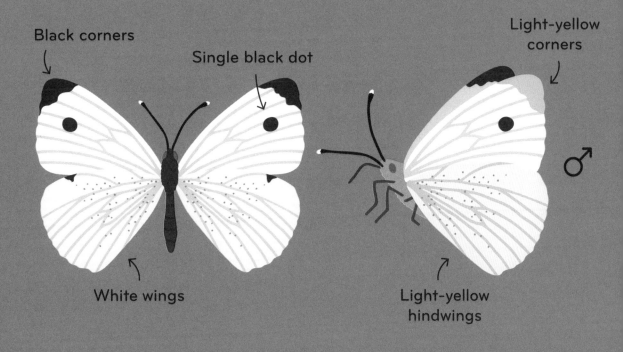

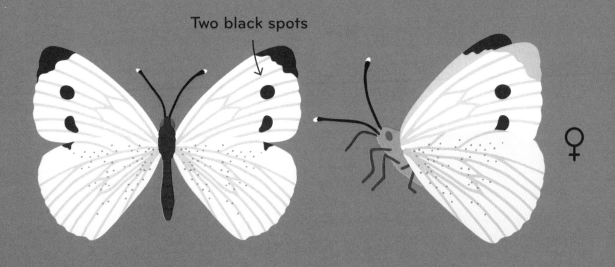

**Where and when to see it**
On flowers and crops, from spring to fall.

**Ease of observation**

# Carolina grasshopper

## The songster

**Order**
Orthoptera

**Activity**
Diurnal

**Habitat**
Countryside

Only male grasshoppers sing. To be more precise, they "stridulate" by rubbing their long hindlegs against their wings. They hear with their abdomen, which is where their eardrums are located!

Grasshoppers are known for their incredible migrations, which they do in swarms. However, the Carolina grasshopper does not migrate and tends to be solitary.

In Asia, grasshoppers are eaten like chips—grilled, salted, and dried—and now people in North America are starting to get in on the fun! Some scientists believe insects could replace meat due to their high levels of protein and lower environmental impact.

**Length**
32 to 58 mm

**Scientific name**
*Dissosteira carolina*

## Metamorphosis

**Egg**
10 days to 2 months

**Larva**
2 to 3 months

**Adult**
2 to 5 months

## Food

The larvae and adults are **phytophagous**, feeding on grasses and seeds.

## Not to be confused with...

The **migratory locust**, which is stockier and has a dark stripe behind each eye and reddish shins.

## Geography

Found in the United States, Canada, and northern Mexico.

## Migration

The Carolina grasshopper is sedentary (it does not migrate).

# Observation guide

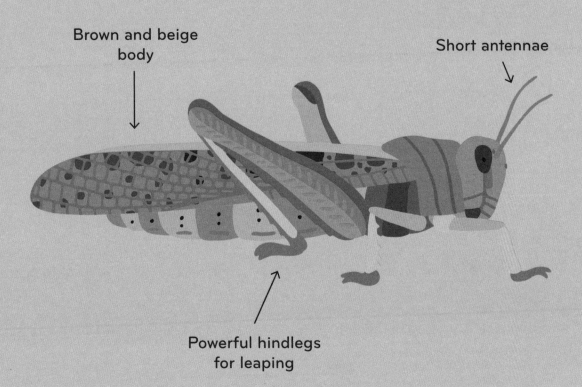

Brown and beige body

Short antennae

Powerful hindlegs for leaping

**Where and when to see it**
In grasses and fields and on gravel roads, mainly during summer.

**Ease of observation**

# A jumping champion

Thanks to their powerful hindlegs, grasshoppers can leap distances of up to 1 meter (3.3 feet)! Sometimes, they spread their wings to gain a little more distance.

# Diamondback moth

## The cabbage eater

**Order**
Lepidoptera

**Activity**
Nocturnal

**Habitat**
Countryside and gardens

The diamondback moth doesn't look like other moths. Its elongated body is only 10 millimeters long and, at rest, its fringed wings fold over its back like a roof. A beige, wavy stripe runs down its body, and its antennae are often pointed forward.

While the adult diamondback moth prefers to feed on flower nectar, the caterpillar has a reputation for destroying crops, such as turnips, mustard, and cabbage. It munches away on the underside of the leaves—the upper surface stays intact—which creates small, see-through patterns. If touched, the caterpillar wriggles and squirms and drops to the ground.

**Length**
~10 mm

**Wingspan**
12 to 17 mm

**Scientific name**
*Plutella xylostella*

## Metamorphosis

**Egg**
3 days

**Caterpillar**
6 to 10 days

**Chrysalis**
4 days

**Adult**
10 days

## Food

The larvae and adults are **phytophagous**. The larvae feed on cabbage, mustard, and turnip leaves, while adults prefer nectar from flowers.

## Not to be confused with...

The **common clothes moth**, which is cream-colored all over. It likes to hide away in houses, whereas the diamondback moth keeps to fields and gardens.

## Geography
Found worldwide.

## Migration
The diamondback moth is sedentary (it does not migrate).

# Observation guide

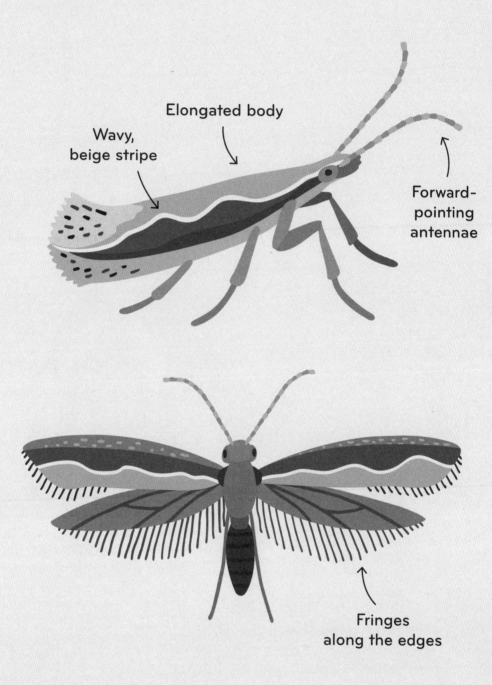

**Where and when to see it**
In vegetable gardens and crop fields, from spring to fall.

**Ease of observation**

# Praying mantis
Praying for prey

**Order**
Mantodea

**Activity**
Diurnal

**Habitat**
Countryside

Hidden among tall grasses, the praying (or European) mantis lies in wait for a crunchy grasshopper or an unfortunate butterfly. This insect has "praying" in its name because its forelegs resemble hands folded in prayer. But these powerful forelegs do more preying than praying—lined with tibial spines, they hook into their victims with astonishing speed! The female is particularly terrifying, with her habit of devouring the male during or after mating. She knows he's a good source of protein for the development of her young.

The praying mantis has a super-sophisticated ability to see the world around it: it can turn its head laterally—something few insects can do—and has 3D vision. Even though it has wings, it's an awkward flyer. It doesn't go after humans, but if you get on the wrong side of one, it will bite to defend itself.

**Length**
Female:
~75 mm

Male:
~50 mm

**Scientific name**
*Mantis religiosa*

**Caution!**
May bite

## Metamorphosis

**Ootheca**
2 to 7 days

**Larva**
2 to 3 weeks

**Adult**
2 to 3 years

## Food

The larvae and adults are **carnivorous**, feeding on grasshoppers, crickets, bees, butterflies, and flies.

## Not to be confused with…

The **Chinese mantis**, which is almost entirely brown and has no spots under its arms.

## Geography
Native to the Mediterranean basin, but also found in Europe, Asia, Africa, and North America, where it is considered invasive.

## Migration
The praying mantis is sedentary (it does not migrate).

# Observation guide

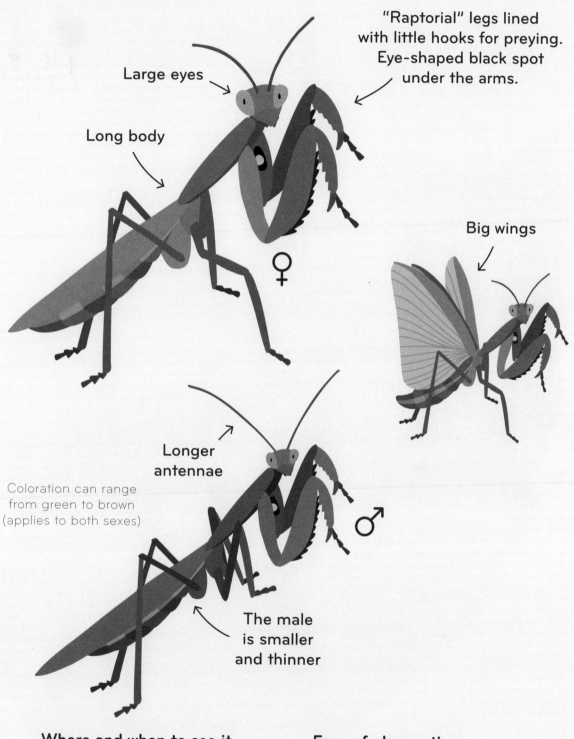

Large eyes

Long body

"Raptorial" legs lined with little hooks for preying. Eye-shaped black spot under the arms.

Big wings

Longer antennae

Coloration can range from green to brown (applies to both sexes)

The male is smaller and thinner

**Where and when to see it**
In tall grasses, dry fields, and vineyards, but only in summer.

**Ease of observation**
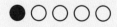

# Fork-tailed bush katydid

Oversized antennae

**Order**
Orthoptera

**Activity**
Crepuscular

**Habitat**
Countryside

During its larval stage, the fork-tailed bush katydid is covered in small black spots, but these start to disappear with age. An expert at mimicry, this insect blends right into surrounding vegetation. It likes to chow down on rosebush and raspberry-cane leaves.

Crickets and grasshoppers can be distinguished from one another by the length of their antennae. Katydids are closely related to crickets, and the antennae of the fork-tailed bush katydid can be as long as its wings! At the end of her abdomen, the female has an "ovipositor"—an organ that enables her to deposit her eggs in tight corners, such as cracks in bark.

The male stridulates by rubbing his wings together and the sound attracts a female. He produces a gentle, high-pitched song, resulting in a summer concerto that starts in the evening and sometimes continues through the night.

**Length**
38 to 45 mm

**Scientific name**
*Scudderia furcata*

## Metamorphosis

**Egg**
8 to 10 months

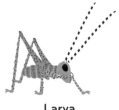
**Larva**
1 to 3 months

**Adult**
1 to 2 months

## Food

The larvae and adults are **phytophagous**, feeding on raspberry, clover, dandelion, and rosebush leaves.

## Not to be confused with...

The **Carolina grasshopper**, which is browner and has shorter antennae. Its wings do not extend as far beyond its abdomen as those of the fork-tailed bush katydid.

## Geography
Found in the United States, Canada, and Mexico.

## Migration
The fork-tailed bush katydid is sedentary (it does not migrate).

# Observation guide

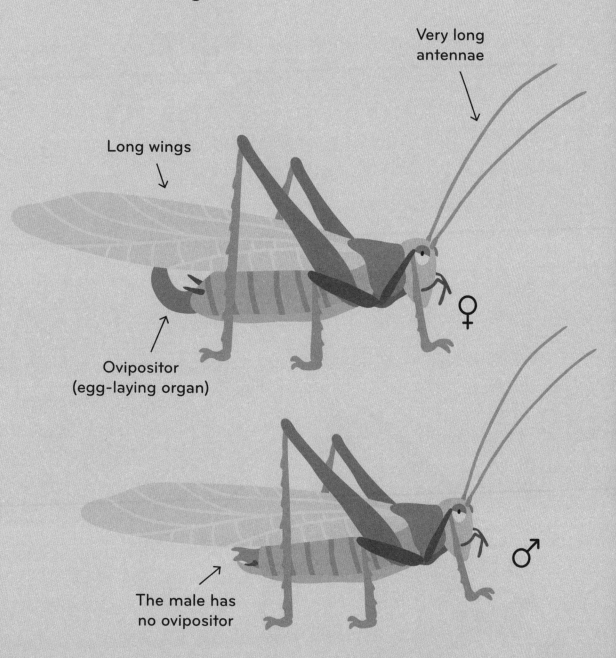

Very long antennae

Long wings

Ovipositor (egg-laying organ) ♀

The male has no ovipositor ♂

## Where and when to see it
In bushes and on leaves, in gardens and parks. Active from early summer to fall.

## Ease of observation

# Monarch

## The magnificent migrator

**Order**
Lepidoptera

**Activity**
Diurnal

**Habitat**
Countryside, towns, cities, gardens, and forests

This large, orange butterfly is famous for its impressive migrations. As summer draws to an end, monarchs prepare themselves for a long journey: from Canada and northeast America to the mountains of Mexico 4,000 kilometers (2,485 miles) away.

The monarch caterpillar has yellow, white, and black stripes. It only eats milkweed leaves, which are almost universally poisonous, but to which it is immune. If a bird is brave enough to eat a monarch caterpillar, however, it will spit it back out immediately!

Sadly, monarchs are an endangered species due to deforestation, industrial farming (which reduces the number of milkweed plants), and climate change.

**Length**
25 to 35 mm

**Wingspan**
70 to 100 mm

**Flying speed**
Up to 30 mph

**Scientific name**
*Danaus plexippus*

## Metamorphosis

**Egg**
1 week

**Caterpillar**
2 to 3 weeks

**Chrysalis**
1 to 2 weeks

**Adult**
2 weeks to 8 months

## Food

The larvae and adults are **phytophagous**. The larvae eat nothing but milkweed leaves, while adults feed on nectar from its flowers.

## Not to be confused with...

The **viceroy**, which is identified by the curved black line that crosses its hindwings,

or the **painted lady**, which has a shorter wingspan (50 mm/2 in.) and a different pattern.

## Geography

Found in North America and the northern areas of South America. They can also be found in southeast Asia, Australia, and New Zealand, and occasionally migrate to western Europe.

## Migration

Some monarchs are migratory.

→ **Learn more about its migration on pp. 212–213**

# Observation guide

Black-veined orange wings

White spots along the wing edge

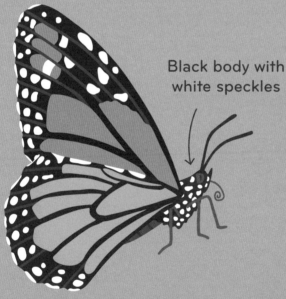

Black body with white speckles

The male has a black dot on the hindwings

**Where and when to see it**
In fields and gardens, from spring to fall.

**Ease of observation**

# Butterfly trees

Monarchs spend the winter in Mexico, where they perch—in the millions—on fir trees. They can cover an entire tree from top to bottom!

# Painted lady

Beautiful and athletic

**Order**
Lepidoptera

**Activity**
Diurnal

**Habitat**
Countryside and forests

The painted lady is one of the most common butterflies in the world. It performs its acrobatics at the edges of forests, in beautiful meadows, and in flowery gardens. This butterfly is migratory and a bit of an athlete: it can travel up to 500 kilometers (310 miles) a day and fly over peaks as high as 2,000 meters (6,500 feet)! In really hot weather, it can even be seen flying upside down in order to protect the delicate scales on its wings. Migrations can take several months, even though the painted lady only lives for a few weeks... In other words, it takes several generations to complete the journey.

The female lays her eggs on host plants such as nettles and thistles. The larvae feed on these plants, which is why the species is sometimes called the thistle butterfly.

**Length**
20 to 30 mm

**Wingspan**
50 to 60 mm

**Flying speed**
Up to 18 mph

**Scientific name**
*Vanessa cardui*

## Metamorphosis

**Egg**
1 week

**Caterpillar**
2 to 6 weeks

**Chrysalis**
1 to 2 weeks

**Adult**
2 weeks

## Food

The larvae and adults are **phytophagous**. The larvae eat nettle, thistle, and lavender leaves, while adults feed on the nectar.

## Not to be confused with...

The **red admiral**, which has black wings with bands of red and orange.

## Geography
Found worldwide.

## Migration
The painted lady is migratory.

→ Learn more about its migration on p. 212

# Observation guide

Orange wings with black and white spots

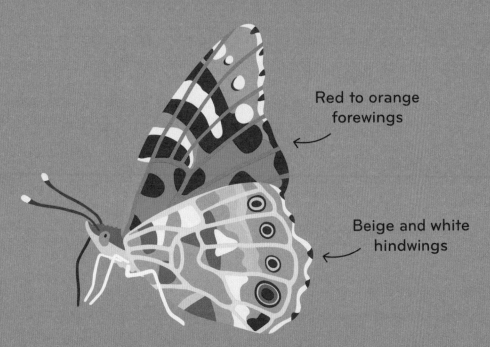

Red to orange forewings

Beige and white hindwings

### Where and when to see it
On flowers, gravel roads, and along forest edges, from early spring to the end of summer.

### Ease of observation

# Small milkweed bug

Red and black keeps predators back

**Order**
Hemiptera

**Activity**
Diurnal

**Habitat**
Countryside, towns, cities, and gardens

This very common bug has a long red and black body. Its thorax and wings have black spots that together form a heart shape.

The small milkweed bug's coloration is a warning to predators—do not eat! Its foul taste comes from ingesting a toxic liquid found in its favorite plant, milkweed. Several other insect species have adopted this appearance too, so that their predators think they're toxic, even though they aren't.

When they mate, the male and female join backsides and can stay in that position for hours. In spring, the female lays her eggs on a milkweed plant so that the larvae have something to eat after they hatch.

**Length**
10 to 12 mm

**Scientific name**
*Lygaeus kalmii*

## Metamorphosis

**Egg**
7 to 10 days

**Larva**
2 to 3 weeks

**Adult**
8 to 12 months

## Food

The larvae and adults are **omnivorous**, feeding on the sap, nectar, and seeds of the milkweed plant, as well as small prey.

## Not to be confused with...

The **large milkweed bug**, which usually has an orangey body. The black on its thorax is shaped more like a diamond than a heart.

## Geography
Found in the United States, Canada, and Mexico.

## Migration
The small milkweed bug is sedentary (it does not migrate).

# Observation guide

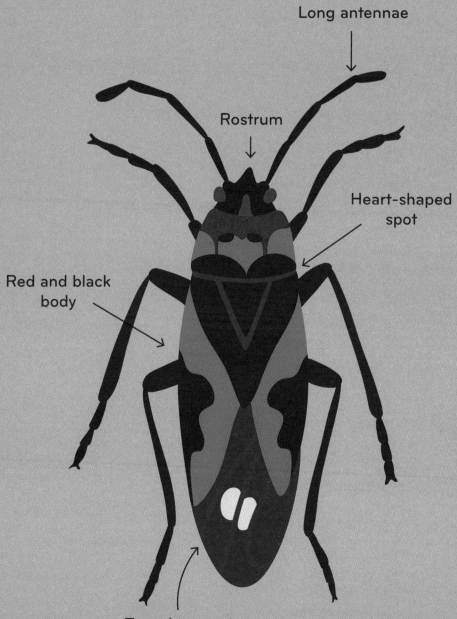

Long antennae

Rostrum

Heart-shaped spot

Red and black body

Two white spots on the membranous wings (sometimes)

**Where and when to see it**
In meadows, on flowers, from spring to fall.

**Ease of observation**

# Spring field cricket

The balcony serenader

**Order**
Orthoptera

**Activity**
Crepuscular

**Habitat**
Countryside and gardens

The spring field cricket spends its life hiding in tall grasses. Using his powerful forelegs, the male digs a 40-centimeter (16-inch) tunnel and builds a nest. The entrance to the hole has a flattened area resembling a balcony. There, in spring, the male takes position and begins to stridulate by rubbing together his elytra, the hardened wings covering his membranous wings. This behavior attracts a mate and marks his territory.

To reproduce, the male slides under the female to deposit his "spermatophore"—a little white sack containing his sperm, which then make their way up to the female's reproductive organs. His mate has an ovipositor—the long, slender organ under her abdomen that allows her to lay eggs in the ground.

**Length**
Female:
14 to 27 mm

Male:
13 to 30 mm

**Scientific name**
*Gryllus veletis*

## Metamorphosis

**Egg**
8 to 12 days

**Larva**
3 to 5 months

**Adult**
2 to 3 months

## Food

The larvae and adults are **omnivorous**, feeding on leaves, seeds, grasses, fruit, and small insects.

## Not to be confused with...

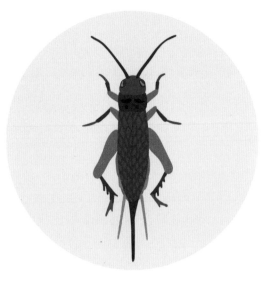

The **house cricket**, which is slimmer and completely golden brown.

## Geography
Found in the United States and Canada.

## Migration
The spring field cricket is sedentary (it does not migrate).

# Observation guide

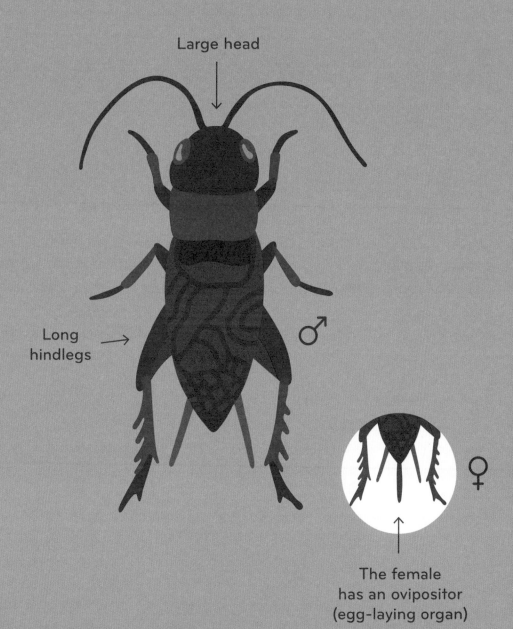

- Large head
- Long hindlegs
- ♂
- ♀ The female has an ovipositor (egg-laying organ)

## Where and when to see it
In tall grasses and fields, as well as on slopes and lawns. Especially active in spring.

## Ease of observation

# Yellow bumble bee

The furball

**Order**
Hymenoptera

**Activity**
Diurnal

**Habitat**
Countryside, towns, cities, and gardens

Plumply shaped and a clumsy flyer, the yellow bumble bee is hardly inconspicuous. Its hindlegs dangle and are equipped with little baskets ("corbiculae") for collecting and carrying pollen. Only females have the ability to sting, which they rarely do, as they prefer to save their energy for visiting pretty flowers full of nectar.

By the end of summer, only the queen remains. In a hollow or abandoned rodent burrow, she prepares a cozy nest lined with pieces of grass, leaves, and hair. Thanks to a reserve of sperm kept for several months in her abdomen, she can start a new colony when sunnier days return. This is a common practice among social members of the Hymenoptera order, including bees and ants.

**Length**
Female worker:
11 to 17 mm

Male:
13 to 16 mm

Queen:
18 to 21 mm

**Scientific name**
*Bombus fervidus*

**Caution!**
May sting

## Metamorphosis

**Egg**
3 to 5 days

**Larva**
7 to 8 days

**Nymph**
12 to 14 days

**Adult**
Female worker/male:
1 month
Queen: 1 year

## Food

The larvae and adults are **phytophagous**, feeding on the nectar of sunflowers, honeysuckle, and clover.

## Not to be confused with...

The **brown-belted bumble bee**, which has a yellow thorax with a black dot in the center. The end of its abdomen is completely black.

## Geography

Found in the United States, Canada, and Mexico.

## Migration

The yellow bumble bee is sedentary (it does not migrate).

# Observation guide

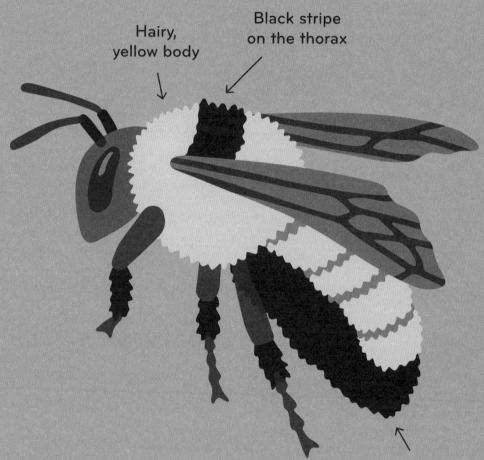

Hairy, yellow body

Black stripe on the thorax

The end and the underside of the abdomen are black

**Where and when to see it**
On flowers, from spring to the end of summer.

**Ease of observation**

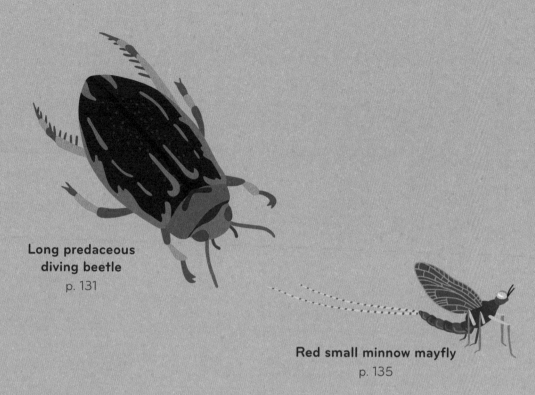

**Long predaceous diving beetle**
p. 131

**Red small minnow mayfly**
p. 135

# Wetlands

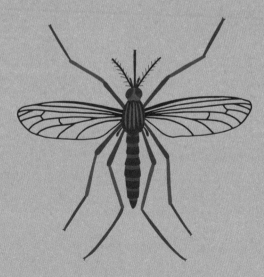

**Northern house mosquito**
p. 139

# Familiar bluet

## The acrobat

**Order**
Odonata

**Activity**
Diurnal

**Habitat**
Wetlands

When resting on a perch, the familiar bluet folds its four delicate wings up over its back. It can be seen around ponds, marshes, and gentle rivers, where it floats from one water lily to another.

When it's time to mate, familiar bluet couples put on quite a performance! The male holds on to an aquatic plant with his legs and grasps the female's head with his cerci. She arches her body until the tip of her abdomen reaches his reproductive organs. The result is a heart-shaped position called a mating wheel. When the act is over, he holds onto her for a little while longer to keep other suitors at bay, and the couple may even remain in this position while flying! Once her eggs are fertilized, the female dives into the water to lay them on plants tens of centimeters deep.

**Length**
28 to 39 mm

**Wingspan**
38 to 44 mm

**Scientific name**
*Enallagma civile*

## Metamorphosis

**Egg**
1 to 3 weeks

**Larva (or naiad)**
2 to 8 months

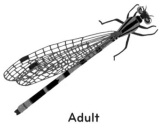
**Adult**
1 to 4 weeks

## Food

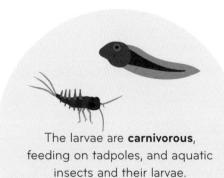

The larvae are **carnivorous**, feeding on tadpoles, and aquatic insects and their larvae.

Adults are also **carnivorous** and eat flies and mosquitoes.

## Not to be confused with…

The **spotted spreadwing**, which has an almost entirely black upper body.

## Geography
Found in North and South America.

## Migration
The familiar bluet is sedentary (it does not migrate).

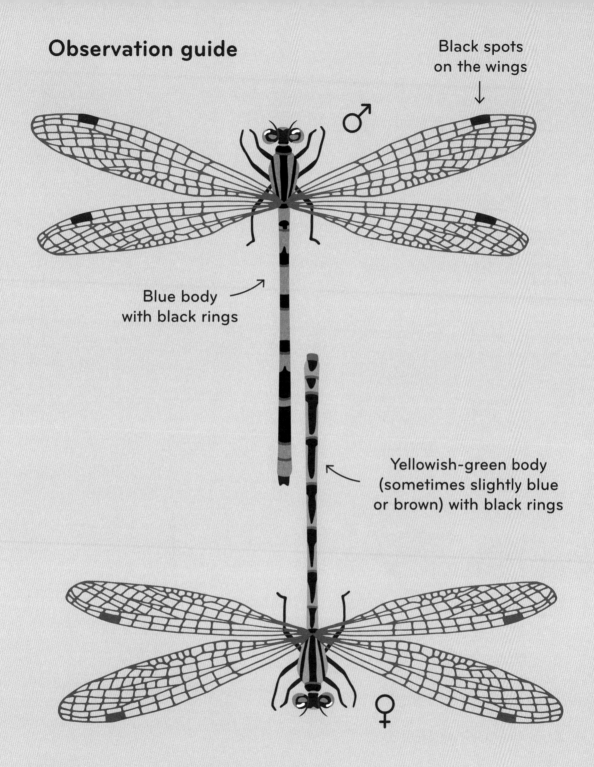

# Dragon or damsel?

The familiar bluet is a member of the Odonata order, but specifically belongs to the suborder of damselflies, not dragonflies. How can you tell them apart?

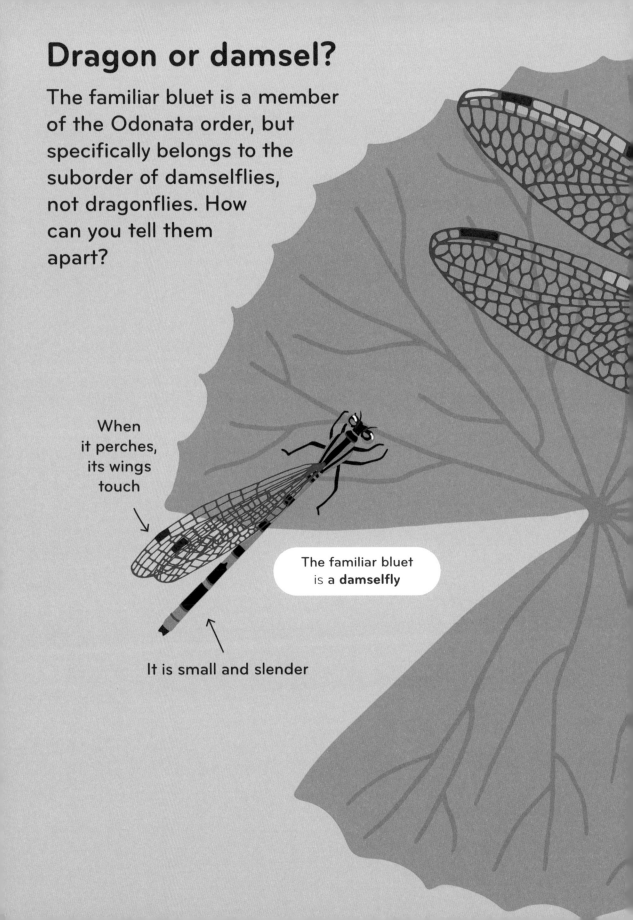

When it perches, its wings touch

The familiar bluet is a **damselfly**

It is small and slender

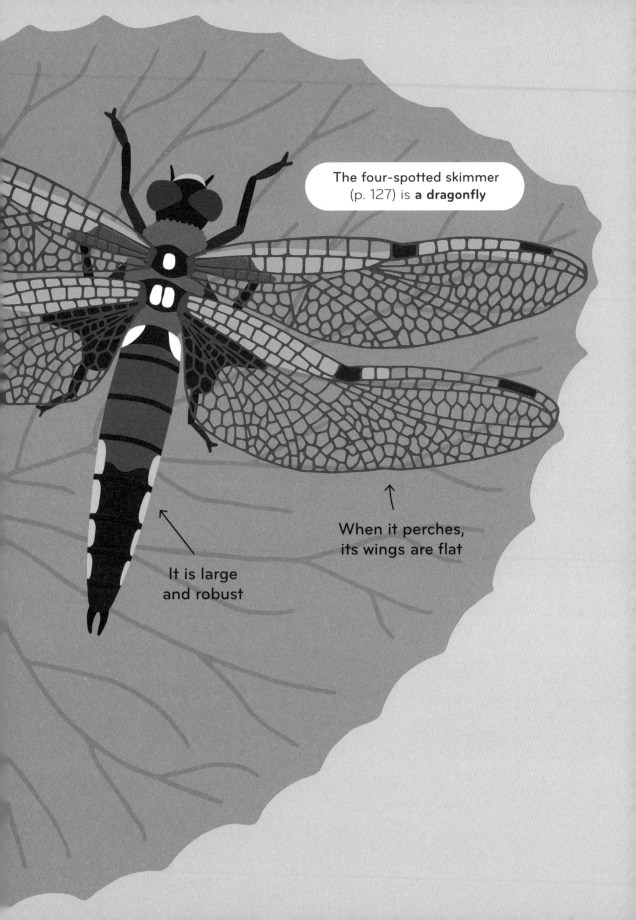

# Four-spotted skimmer

A flying gem

**Order**
Odonata

**Activity**
Diurnal

**Habitat**
Wetlands

When perched atop a tree branch, the four-spotted skimmer sparkles and shines. This dragonfly stands out thanks to its golden-brown body and four wings edged with black spots. The male is known to be aggressive towards other insects when defending his marshy territory.

Mating couples reproduce in flight, and the whole act takes just a few seconds. The female then deposits her eggs on floating vegetation, and it takes just a day for the larvae to emerge. The larvae spend a few weeks in the water, where they molt several times before reappearing as adults in all their glory.

As adults, these large dragonflies feast on flies and mosquitoes, while the larvae eat tadpoles.

**Length**
40 to 50 mm

**Wingspan**
60 to 80 mm

**Scientific name**
*Libellula quadrimaculata*

## Metamorphosis

**Egg**
1 day

**Larva (or naiad)**
2 to 3 weeks

**Adult**
1 to 2 months

## Food

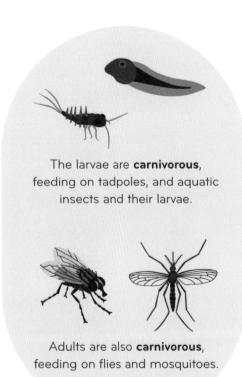

The larvae are **carnivorous**, feeding on tadpoles, and aquatic insects and their larvae.

Adults are also **carnivorous**, feeding on flies and mosquitoes.

## Not to be confused with…

The **wandering glider**, which has a yellow, more delicate body. Each wing has a single red spot.

## Geography
Found in Europe, Asia, and North America.

## Migration
The four-spotted skimmer is migratory.

# Observation guide

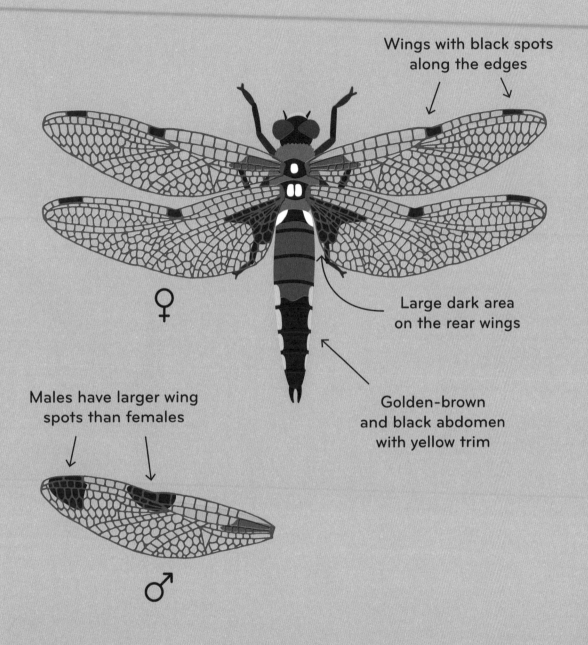

Wings with black spots along the edges

Large dark area on the rear wings

Golden-brown and black abdomen with yellow trim

♀

Males have larger wing spots than females

♂

### Where and when to see it
On branches and reeds in wetlands and along small mountain lakes, from spring to summer.

### Ease of observation

# Long predaceous diving beetle

## The free diver

**Order**
Coleoptera

**Activity**
Diurnal

**Habitat**
Wetlands

Stagnant waters with abundant vegetation are where you'll find this amazing water beetle. By storing oxygen under its elytra, the long predaceous diving beetle can stay underwater for several minutes. Its hindlegs are flat like tiny paddles, allowing it to swim around in search of food and to get away from hungry herons!

In early summer, the larvae change into nymphs. They leave the water to bury themselves in the ground, then emerge about two weeks later as adults. At night, adults often fly to many water sources in search of a mate.

If you happen upon one of these insects, it's best to leave it alone because they are very good at defending themselves. Adults are equipped with very sharp spurs on their shins, while the larvae defend themselves with their jaws, which are capable of inflicting a bite as painful as a wasp sting.

**Length**
7 to 10 mm

**Scientific name**
*Coptotomus longulus*

**Caution!**
Larvae bite
and adults can cut
with their hindlegs

## Metamorphosis

**Egg**
7 to 10 days

**Larva**
2 to 4 weeks

**Nymph**
1 to 2 weeks

**Adult**
2 years

## Food

The larvae and adults are **carnivorous**, feeding on small fish, tadpoles, and fry. Adults sometimes eat the larvae of other aquatic insects.

## Not to be confused with...

The **giant water scavenger beetle**, which is entirely black and can grow to between 32 and 40 millimeters in length.

## Geography

Found in the United States and southern parts of Canada.

## Migration

The long predaceous diving beetle is sedentary (it does not migrate).

# Observation guide

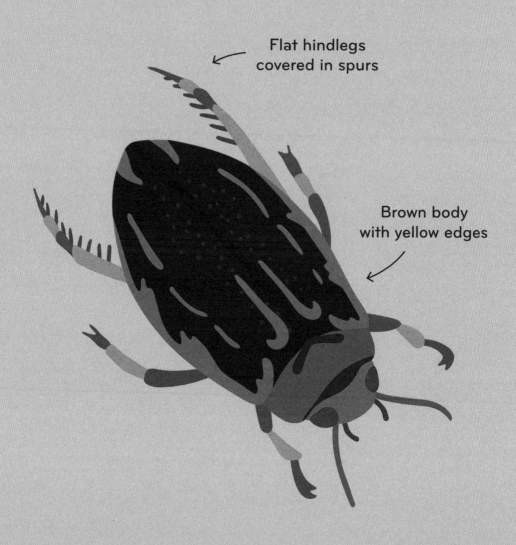

Flat hindlegs covered in spurs

Brown body with yellow edges

### Where and when to see it
In wetlands, especially swamps and marshes. This beetle is difficult to observe as it spends most of its time underwater. Active in spring and summer.

### Ease of observation
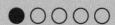

# Red small minnow mayfly

Here today, gone tomorrow

 |  |

**Order**
Ephemeroptera

**Activity**
Diurnal

**Habitat**
Wetlands

Mayflies belong to the Ephemeroptera order, and are indeed "ephemeral," meaning short-lived: adults live anywhere from a few hours to a few days. During the adult stage, a mayfly's mouth no longer works, so it cannot eat. The sole purpose of its short life is to reproduce, as evidenced by the clouds of males and females hovering in a courtship dance! After mating in flight, the male quickly dies. The female finds a water surface, where she'll lay thousands of eggs before it's her turn to die. The aquatic larvae live for two or three weeks.

At rest, the red small minnow mayfly holds its wings upright and they touch. At the tip of its abdomen are two tail-like appendages called "cerci," which help the insect remain stable in flight.

Lots of different animals eat mayflies, including birds, bats, and fish.

**Length**
10 to 14 mm (including the cerci)

**Wingspan**
~20 mm

**Scientific name**
*Callibaetis ferrugineus*

## Metamorphosis

**Egg**
2 to 7 days

**Larva**
2 to 3 weeks

**Adult**
2 to 3 days

## Food

The larvae are **saprophagous**, feeding on tiny, decomposing algae. Adults do not feed.

## Not to be confused with...

The **Michigan hex burrowing mayfly**, which is larger (~25 mm) and usually yellow and brown. Its shape is less arched, and its forelegs are often stretched out in front of it. It has two pairs of visible wings.

## Geography

Found in Canada, the United States, and northern Mexico.

## Migration

The red small minnow mayfly is sedentary (it does not migrate).

# Observation guide

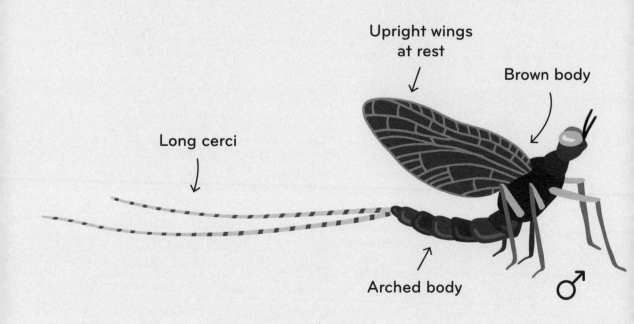

Upright wings at rest
Brown body
Long cerci
Arched body
♂

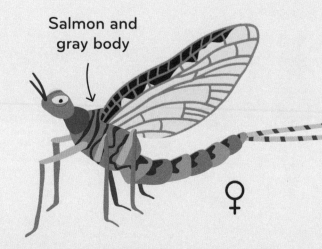

Salmon and gray body
♀

**Where and when to see it**
In wetlands and on windowpanes, especially in summer.

**Ease of observation**

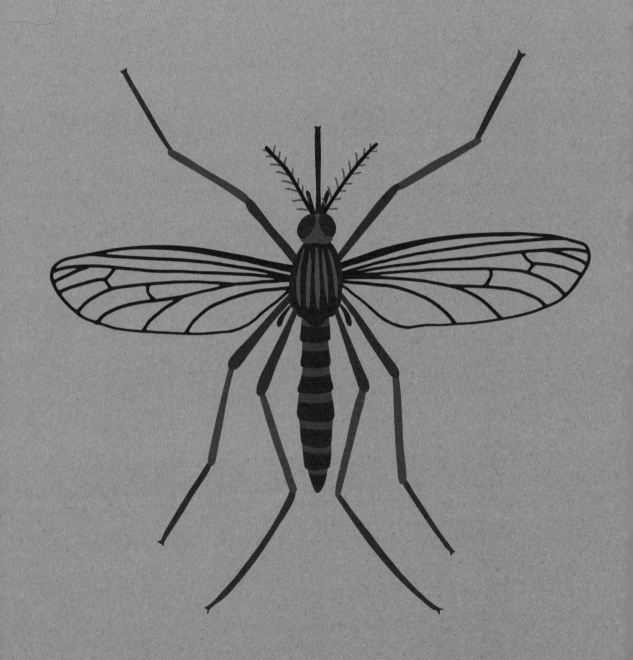

# Northern house mosquito

Small but noisy

**Order**
Diptera

**Activity**
Crepuscular

**Habitat**
Wetlands, forests, and towns and cities

The buzzing sound that mosquitoes make—the one that can keep us awake at night—comes from the rapid beating of the female's wings. When a female bites us with her proboscis, it's for the sole purpose of drawing blood to feed her eggs. Her saliva contains an anesthetic, which is why we don't feel her bite. When our skin swells and itches, it's just our immune system reacting to her venom.

The male mosquito has antennae that are quite feathery, and these allow him to detect female pheromones from 10 feet away. He, too, has a proboscis, but he only uses it to gather nectar from flowers or sap from trees. The female mosquito lays her eggs on the surface of standing water.

**Length**
5 to 6 mm

**Wingspan**
5 to 7 mm

**Scientific name**
*Culex pipiens*

**Caution!**
May bite

## Metamorphosis

**Egg**
1 to 2 days

**Larva**
1 week

**Nymph (or pupa)**
1 to 2 days

**Adult**
2 days to 2 months

## Food

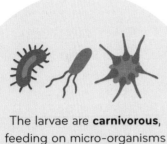

The larvae are **carnivorous**, feeding on micro-organisms in the water.

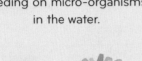

Adult females are also **carnivorous**, as they drink blood. Males prefer nectar and sap.

## Not to be confused with...

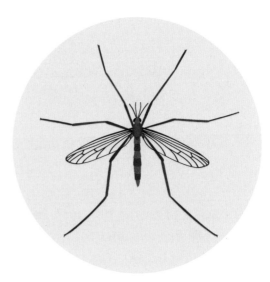

The **common crane fly**, also known as daddy long legs, which is much bigger (~25 mm) and does not bite.

## Geography
Found worldwide.

## Migration
The northern house mosquito is sedentary (it does not migrate).

# Observation guide

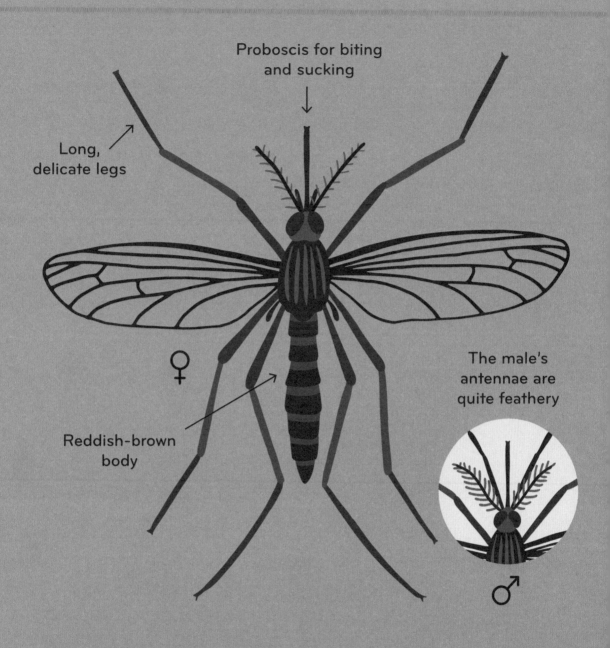

Proboscis for biting and sucking ↓

Long, delicate legs

♀ Reddish-brown body

The male's antennae are quite feathery

♂

### Where and when to see it
On windows, around houses, and near standing water. Active from spring to fall but more commonly found near homes during summer.

### Ease of observation

**Eastern subterranean termite**
p. 145

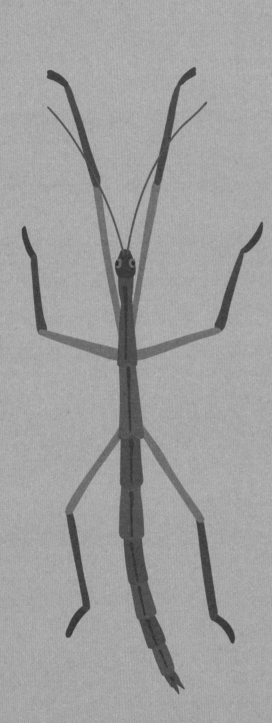

**Northern walkingstick**
p. 149

**Odorous house ant**
p. 155

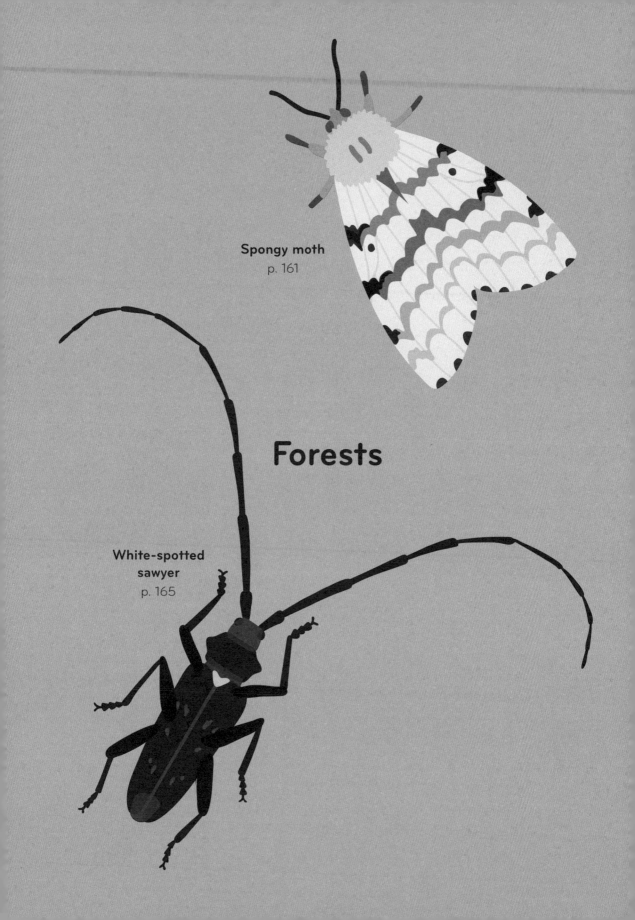

# Forests

**Spongy moth**
p. 161

**White-spotted sawyer**
p. 165

# Eastern subterranean termite

A team player

**Order**
Blattodea

**Activity**
Nocturnal

**Habitat**
Forests

Eastern subterranean termites are social insects that live in the forest but sometimes turn up in our houses, where they can damage the wood. They form well-organized colonies: sterile (or asexual) workers for gathering food and tending to the nest; soldiers for protecting the colony; and reproductive (or sexual) termites that leave the nest in the fall to start new colonies. Colonies are governed by a royal couple—a king and queen responsible for reproduction. If one of them dies, a worker steps up as a replacement and becomes sexual!

There are numerous termite species in the world. The largest populations are found in tropical regions, where they build nests from chewed-up earth. These mounds can be up to 10 feet tall.

**Length**
Worker:
3 to 4 mm

Soldier:
6 to 7 mm

Reproductive individual and the royal couple:
8 to 10 mm

**Scientific name**
*Reticulitermes flavipes*

## Metamorphosis

**Egg**
2 to 7 weeks

**Larva**
2 to 5 weeks

**Reproductive adult**
2 to 3 years

## Food

The larvae and adults are **xylophagous**, feeding on the pulp of dead wood.

## Not to be confused with...

The **odorous house ant** (p. 155), which is larger and has antennae that are bent at an angle.

## Geography
Found in the United States, Canada, and Mexico.

## Migration
The eastern subterranean termite is sedentary (it does not migrate).

# Observation guide

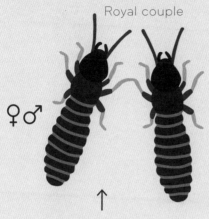

**Royal couple**
♀ ♂
↑
Brown bodies with yellow stripes

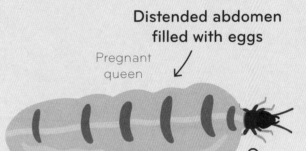

**Pregnant queen**
**Distended abdomen filled with eggs**
♀

**Soldier**

↑
Light-yellow body and large head with two large jaws

**Worker**

↑
White body

**Reproductive termite**

↑
Brown body with wings

## Where and when to see it
In forests, on stumps and dead wood. Especially active in summer.

## Ease of observation
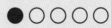

# Northern walkingstick

The phantom

**Order**
Phasmatodea

**Activity**
Nocturnal

**Habitat**
Forest

The word phantom comes from the Greek "phasma," meaning ghost or phantom, and insects in the Phasmatodea order are often called "phasmids." The northern walkingstick is one such insect and it behaves somewhat like a phantom—because it looks a lot like a stick, it's practically invisible among the leaves! Plus, it tends to be more active at night, making it even harder to see. This insect is normally very still, but when it must move, it does so by swaying back and forth to mimic the way branches move in the wind.

Birds unintentionally play a role in the reproduction of phasmids: they eat their eggs but do not digest them. Instead, the eggs pass through the birds' digestive systems intact and are then released in their droppings.

Males are very rare, and even altogether absent in certain phasmid species. In these cases, females reproduce all by themselves through parthenogenesis, like aphids and roaches.

**Length**
Female:
70 to 100 mm

Male:
68 to 84 mm

**Scientific name**
*Diapheromera femorata*

## Metamorphosis

**Egg**
2 weeks
to 5 months

**Larva**
2 to 4 months

**Adult**
1 to 3 months

## Food

The larvae and adults are **phytophagous**, feeding on the leaves of brambles, rosebushes, and cherry and oak trees.

## Not to be confused with...

The **giant walkingstick**, which is larger (it grows up to 150 millimeters (almost 6 inches) long) and much more colorful.

## Geography
Found in the United States, Canada, and Mexico.

## Migration
The northern walkingstick is sedentary (it does not migrate).

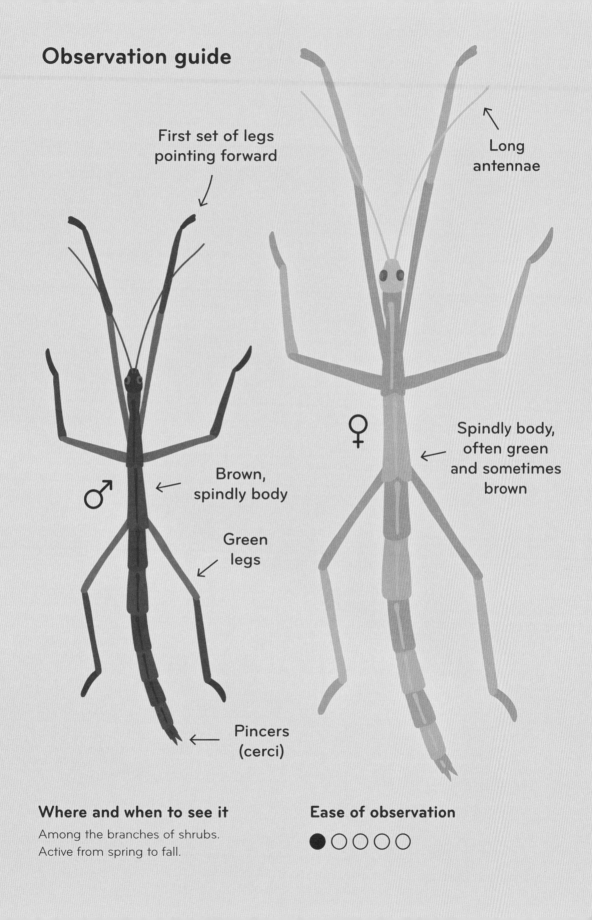

# The art of camouflage

Insects in the Phasmatodea order, sometimes called ghost insects or phasmids, come in strange shapes. Some resemble twigs, while others look like leaves or pieces of bark. Some can even change their color to match their surroundings. These species are most commonly found in Australia, western Asia, and Indonesia.

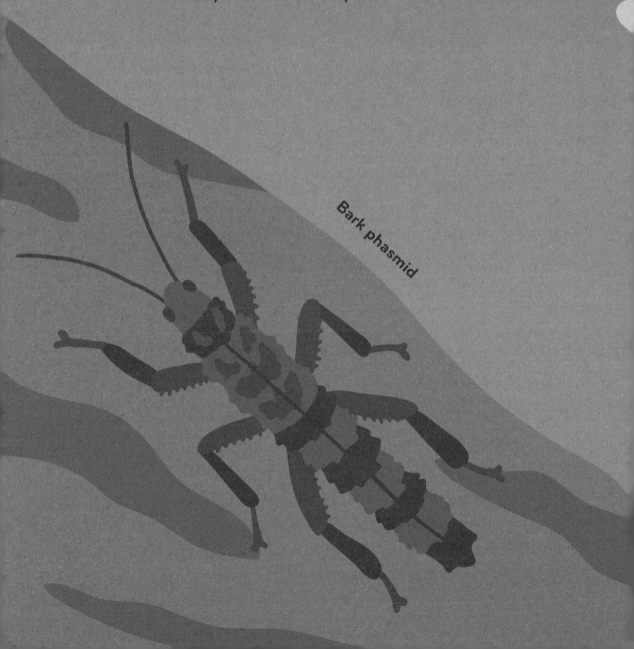

Bark phasmid

# Odorous house ant

The mighty one

**Order**
Hymenoptera

**Activity**
Diurnal

**Habitat**
Forests, countryside, towns, cities, and gardens

The odorous house ant is perfectly at home in any habitat: gardens, forests, and cracks and crevices around the house—it will even waltz right into the kitchen in search of something sweet. This species of ant builds well-organized nests out of twigs, needles, and dead leaves. They're called odorous house ants because they smell like rotten coconut or blue cheese when stepped on!

There are 12,000 species of ant in the world. Ants are resilient social insects that conquered every continent during the Jurassic period, 100 million years ago. They are capable of lifting 10 to 20 times their body weight! Sometimes they raise colonies of aphids so they can feed on the sweet, nourishing honeydew they secrete. Ants are small but most can protect themselves very effectively—they spray formic acid from the tip of their abdomen, which predators find dreadfully itchy. The acid poses no risk to humans.

**Length**
Female worker: 2 to 3 mm

Queen and male: 4 to 5 mm

**Scientific name**
*Tapinoma sessile*

**Caution!**
May bite (painless) and project formic acid

## Metamorphosis

**Egg**
2 to 3 weeks

**Larva**
2 to 4 weeks

**Nymph**
2 to 3 weeks

**Adult**
Female worker: A few months
Male: 1 week
Queen: 8 to 12 months

## Food

The larvae and adults are **omnivorous**, feeding on whatever they find, but mainly dead insects and the honeydew produced by aphids.

## Not to be confused with...

The **pharaoh ant**, which is orange and brown, but smaller (1 to 2 mm),

or the **black carpenter ant**, which is larger (6 to 16 mm).

## Geography
Found in the United States, Canada, and Mexico.

## Migration
The odorous house ant is sedentary (it does not migrate).

# Observation guide

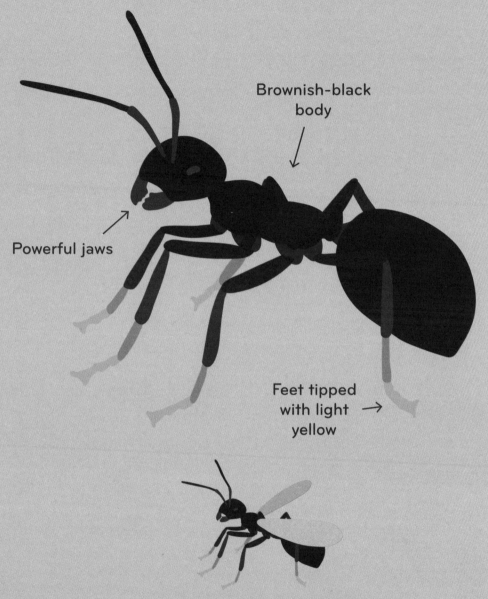

Brownish-black body

Powerful jaws

Feet tipped with light yellow →

Reproductive ants have wings during mating season

### Where and when to see it
In forests, under rocks, and inside tree stumps and houses. Especially active in spring and summer.

### Ease of observation

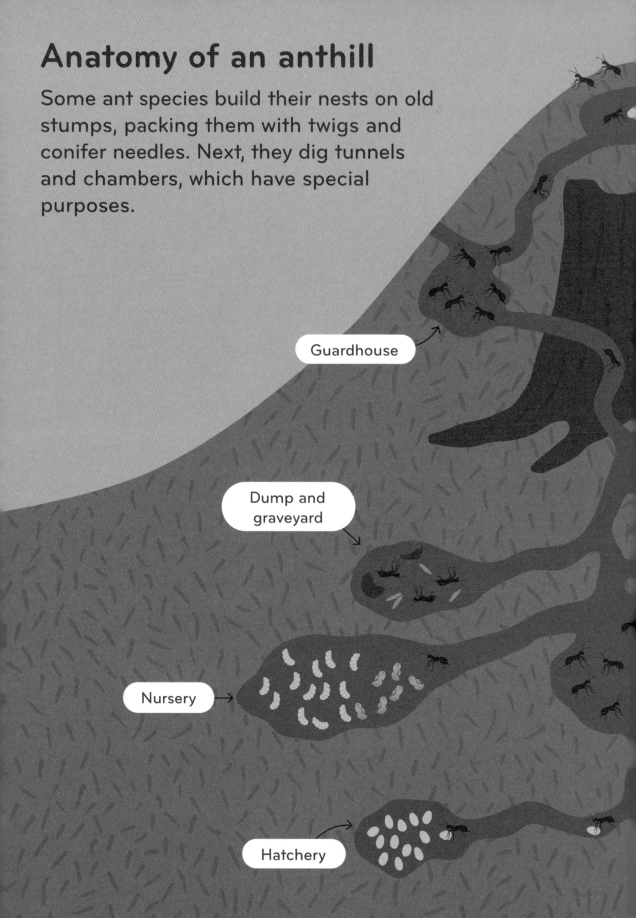

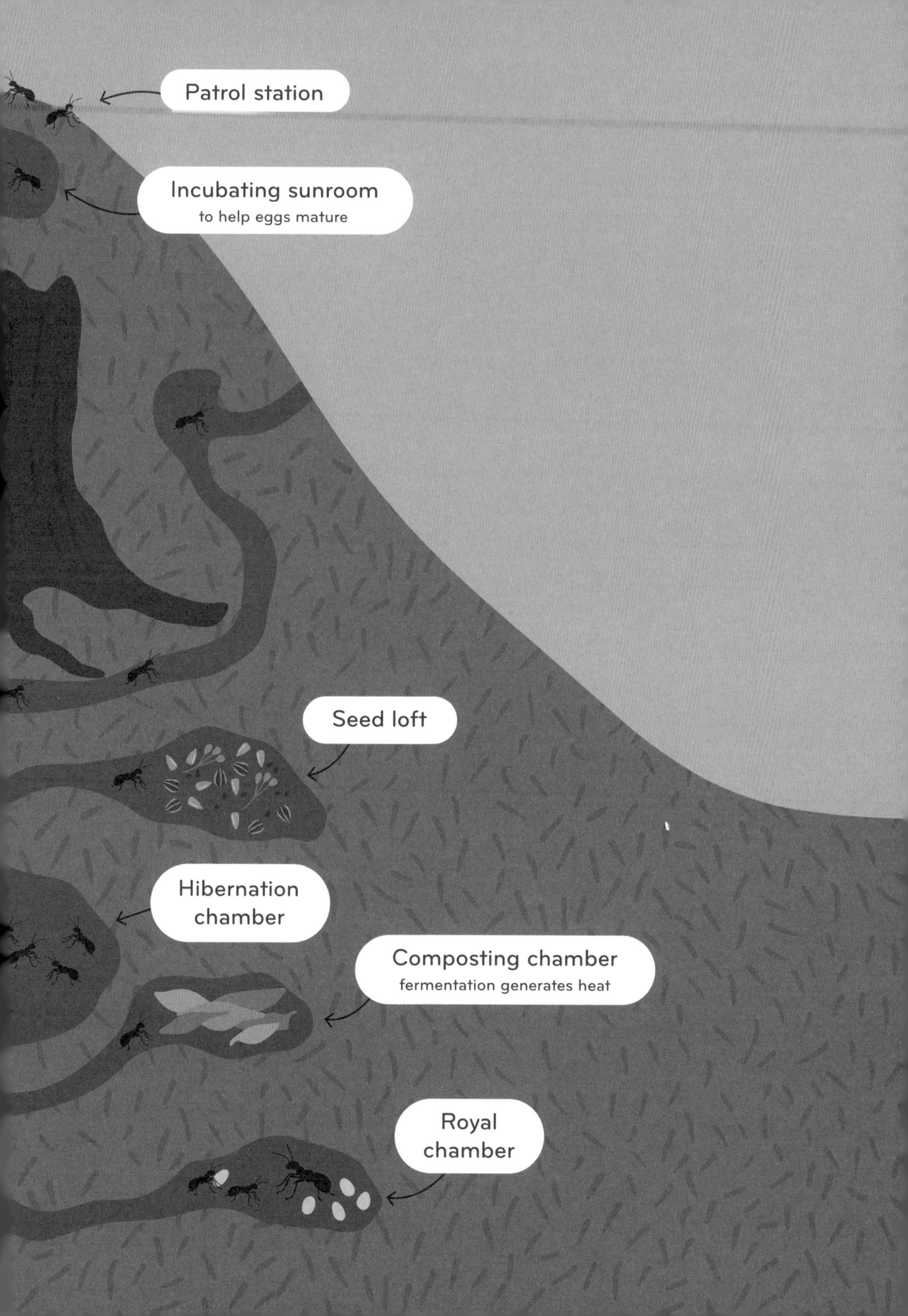

- Patrol station
- Incubating sunroom
  to help eggs mature
- Seed loft
- Hibernation chamber
- Composting chamber
  fermentation generates heat
- Royal chamber

# Spongy moth

The tree eater

**Order**
Lepidoptera

**Activity**
Nocturnal

**Habitat**
Forests, towns, and cities

When summer temperatures soar, this insect thrives. Spongy moth caterpillars have a voracious appetite, as well as a bad reputation for ravaging chestnut, birch, and oak trees. They can decimate many acres of forest in a single season! The good news is that the droppings they leave behind on dead leaves cover the forest floor with an effective fertilizer.

The spongy moth is exceptionally good at camouflaging itself. Thanks to the color of its wings, it is almost invisible against tree bark, which protects it from robins and other birds.

The female is creamy white, while the male is reddish-brown and sports two attractive feathery antennae on his head for detecting female pheromones.

**Length**
15 to 30 mm

**Wingspan**
30 to 70 mm

**Scientific name**
*Lymantria dispar*

**Caution!**
The caterpillar has prickly hairs that cause itchy rashes.

## Metamorphosis

**Egg**
3 to 4 months

**Caterpillar**
2 to 3 months

**Chrysalis**
2 weeks

**Adult**
4 to 9 days

## Food

The larvae are **phytophagous**, feeding on leaves and needles. Adults do not eat.

## Not to be confused with...

The **salt marsh moth**, which has white wings with black specks. The male has orange hindwings.

## Geography

Can be seen in Europe, northern Africa, central Asia, and North America, where it is considered invasive.

## Migration

The spongy moth is sedentary (it does not migrate).

# Observation guide

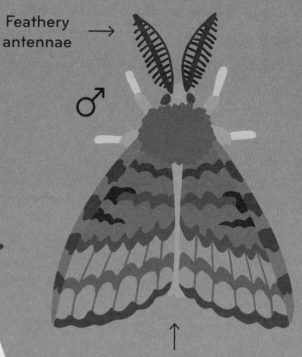

Feathery antennae →

♂

↑
Reddish-brown body

♀

↑
Creamy white body with streaks of beige and black

**Where and when to see it**
On tree trunks on summer nights.

**Ease of observation**

# White-spotted sawyer

## The conifer dweller

**Order**
Coleoptera

**Activity**
Diurnal

**Habitat**
Forests

The white-spotted sawyer is a long-bodied beetle. It is easily recognizable by its sizable antennae, which are as long as the female's body and twice as long as the male's. A white, heart-shaped spot can be found on the nape of its neck. During flight, the white-spotted sawyer's body is almost vertical.

This insect lives in coniferous forests—pine, fir, spruce, and larch make good homes and provide an excellent source of food. The female makes tiny cracks in the bark to lay her eggs. Once hatched, the larvae nibble away at the wood, digging tunnels up to 15 centimeters (6 inches) in length. White-spotted sawyer larvae generally go for sick or decomposing trees, thus helping with forest renewal. As for adults, they prefer to chow down on needles and young twigs.

**Length**
18 to 25 mm (excluding the antennae)

**Scientific name**
*Monochamus scutellatus*

## Metamorphosis

**Egg**
2 to 8 weeks

**Larva**
1 year

**Nymph**
4 to 6 weeks

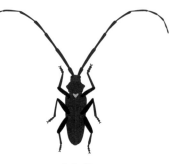
**Adult**
2 months

## Food

The larvae are **xylophagous**, feeding on dead wood. Adults are **phytophagous**, feeding on needles and young twigs.

## Not to be confused with...

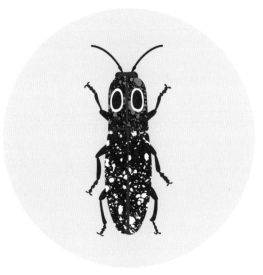

The **eyed click beetle**, which has shorter antennae and two spots resembling black eyes on its thorax.

## Geography
Found in eastern and western North America and in Alaska.

## Migration
The white-spotted sawyer is sedentary (it does not migrate).

# Observation guide

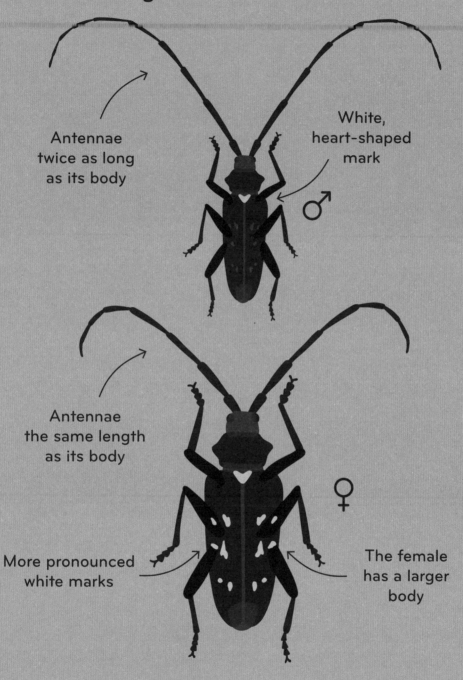

Antennae twice as long as its body

White, heart-shaped mark

♂

Antennae the same length as its body

More pronounced white marks

The female has a larger body

♀

**Where and when to see it**
On conifer trunks, on fine summer days.

**Ease of observation**

# Other astonishing insects

# A herculean size

The Hercules beetle is 17 centimeters (6.7 inches) long, making it one of the largest in the world!

Found in Central and South America

# The water walker

There are insects in the Hemiptera order, called water striders, skimmers, scooters, or skaters, that glide on the water's surface. They do this by spreading and angling their legs, which are equipped with tiny water-repellent hairs.

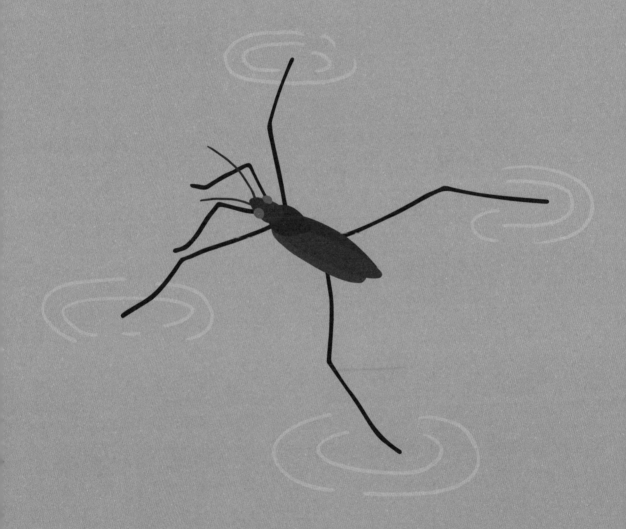

Found worldwide

# A long-burrowing nymph

Cicadas are known for their singing, but it's their longevity that's truly impressive! As nymphs, periodical cicadas can take 17 years to mature underground!

Found in North America

# The intimidator

The forest giant owl is a large butterfly with a 12-centimeter (4.7-inch) wingspan. Its wings are adorned with large black and yellow circles, which look like frightening owl eyes to would-be predators.

Found in Central and South America

# King of the pond

The green darner is one of the world's largest flying insects, measuring 7 centimeters (2.7 inches) in length and 8 centimeters (3.1 inches) across the wings. This large dragonfly defends its territory by patrolling pond surfaces. Its royal status is reflected in its scientific name, *Anax junius* —"Anax" means king in Ancient Greek.

Found in Central and North America

Found in North
and Central America

# The impostor

*Myrmeleon immaculatus* is a species of antlion that closely resembles a dragonfly, but with longer antennae. It is in the Neuroptera order of net-winged insects.

# A bright-light butt!

Fireflies produce light signals from the tips of their abdomens to identify other members of their species and attract a mate. This phenomenon, called "bioluminescence," is a reaction that converts chemical energy into light energy.

Found worldwide
(except in polar regions)

Found in North America, Europe, and Asia

# The scorpion fly

The scorpion fly is in the Mecoptera order and has quite the impressive physique. Despite being just 2 centimeters (0.8 inches) long, it has an elongated snout, and the male is equipped with a claw-shaped reproductive organ at the end of its abdomen that resembles a scorpion's stinger.

# A hummingbird look-alike

The snowberry clearwing is a species of moth that is sometimes mistaken for a tiny hummingbird! It has a long proboscis for drawing nectar from deep within flowers.

Found in Central and North America

# 2

# Go on, take a closer look!

Where do insects hide out? What do they do during winter? What are our best chances of seeing them? Insects have lots to teach us, so open your eyes and be patient.

# Tips for getting started

### ① Start with the easy ones

Ants, honey bees, bumble bees, butterflies, and ladybugs are some of the most popular insects, and with good reason—they're easy to find! This adventure can start in your own backyard or a neighborhood park. Spring and summer are the best seasons for insect watching.

### ② Patience and perseverance

Insect watching requires patience and perseverance because their world is much smaller than ours. You have to get right up close to bushes and trees, turn over stones, and crouch down on the ground. If today didn't turn up much, don't get discouraged, because tomorrow might be full of wonderful discoveries!

### ③ A brain game

The more you learn about insects, the easier it will be to see and recognize them! Memorize their shapes and colors, look at pictures of them in books and online, and watch videos and documentaries about them.

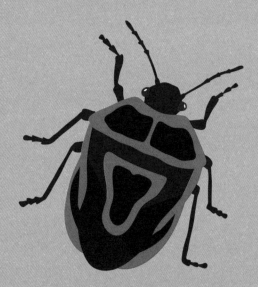

Two-spotted stink bug

# How to identify an insect

① **Note its appearance**

- What color(s) is it? Does it have spots or stripes?
- Are its antennae long or short?
- Based on its appearance, are you able to determine what order it belongs to?

② **Note its habitat**

- Did you see it near water? Or in a park or forest?
- Was it in a tree or in tall grass?
- Did you find it in your house?

③ **Note its behavior**

- Was it alone or in a group?
- Did it move by flying, jumping, or walking?
- Was it feeding or cleaning itself when you found it?

Try to find it in this handbook or look online with search words like *insect + red + stripes + the name of your country.*

Ichneumon centrator

# Get the right gear

**Your *Insectorama* handbook**

**A notepad and pencil**
for jotting down your observations

**A hat and sunglasses**

**Good shoes**

**Mosquito repellent
or anti-tick spray**

**Food and water**

# Get a better look

Insects live life on a miniature scale, but there are tools you can use to get a good look at them.

**A camera or binoculars**

**A magnifying glass**

**A flashlight**
for insect watching at night

# Insect collecting

There are lots of tools you can use to capture and observe insects, but it's always best not to touch them at all. Leaving them undisturbed will spare them from unnecessary stress and accidental injuries. Keep in mind that you need permission to capture insects in some places.

**Tweezers**

**A magnifying box**

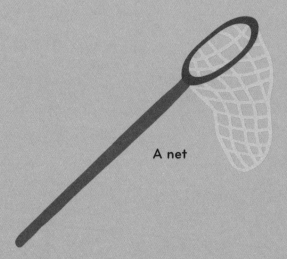
**A net**

**A beating sheet**
for catching falling insects

# Rolling with the seasons

## Spring

The snow has melted and the first flowers are blooming. By March, bees and flies are already busy pollinating. Eggs, larvae, and nymphs have made it through the winter and are starting the next stage of their development.

## Summer

There's no doubt about it—summer is the best time for insect watching! Butterflies float from one flower to another, beetles tramp through the bushes, and caterpillars munch on lush green leaves. Grasshoppers and crickets sing the sun to sleep.

## Fall

The flowers have withered and, slowly, bees take their final bow. Ripe fruit falls from the trees to rot in the grass, providing juice for hungry insects. Evenings get cooler, and insects get ready for winter.

## Winter

### They hide

Ladybugs, lacewings, and bugs seek shelter in houses, while earwigs slip under the bark of trees. They spend the winter as **adults** and their bodies enter a state of dormancy known as **diapause**.

### They change form

Female mantises, grasshoppers, and crickets lay their eggs in fall before dying. Only the **eggs** survive winter, turning into larvae in spring. Many beetles spend the winter as **larvae** buried underground. As for butterflies, they're often **chrysalides** during winter.

### They migrate

Many insects fly away to warmer places. Butterflies and dragonflies migrate the greatest distances.

### They die

There are some species of wasps and bees (the bumble bee, for instance) where the whole colony dies—except for the queen. She will dip into her reserve of sperm to start a new family when the warm weather returns.

# Towns, cities, and gardens

- In trees
- On windows
- In hedges
- On flowers
- In vegetable gardens
- In the ground
- In bushes

# Countryside

In flowery meadows

On gravel roads

In crop fields

# Forests

On leaves

In cut wood

On or in bark

Under dead leaves

In old stumps

Under rocks

# Who goes there?

### A punctured pecan?
Perhaps a pecan weevil has laid its eggs in the hole.

### A chewed-up leaf?
Maybe the leafcutter, a species of wild bee, cut out pieces to line its nest.

### A tree trunk with patterns on it?
It might be the work of the larvae of the European elm bark beetle!

### A leaf with bite marks?
A sure sign a caterpillar came by to fill up.

### Strange growths?
Perhaps it's a gall wasp, a small member of the Hymenoptera order that lays its eggs on oak leaves and causes galls (abnormal growths) to form.

### A gnawed-up old stump?
No doubt it's a sign of larvae that feed on dead wood, like those of the white-spotted sawyer.

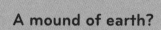

### A mound of earth?
Some species of ants dig tunnels in the ground, tossing out soil through holes and creating little piles of dry earth.

# Why are some people afraid of insects?

Even though small creatures generally don't eat big ones, we still sometimes find them scary. Why?

## A natural reflex

Fear is a basic part of survival. Humans, like all animals, are constantly assessing danger levels to keep ourselves safe. Fear is a physical response to potential danger, and it can be triggered even when the insect before us is harmless. The mere sight of the insect can awaken impulses encoded in our DNA.

## Cultural influence

Movies, tales, and legends play an important role in shaping our imagination, including the way we react to insects. Cockroaches are associated with filth, worms conjure up images of death, and spiders invoke terror (whereas ladybugs symbolize luck, and butterflies elegance!). Some insects that are often scorned in the Western World are revered in other regions; for instance, ancient Egyptian pharaohs wore jewelry shaped like holy beetles, and spiders represent wisdom in many African traditions.

## Take a step back

We tend to fear what we don't know, but if we took the time to familiarize ourselves with insect names, behaviors, and traits, we might just learn to like them!

So, when you see an insect that scares you, first ask yourself if it really presents a danger. If you know it doesn't, tell yourself that it's just going about its business and has absolutely no intention of hurting you!

3

# The life of insects

How do insects communicate? Why do they migrate? What role do they play in the ecosystem? Whether they're larvae or adults, these little animals have a very busy life and are the backbone of biodiversity.

# Magical metamorphoses

Insects undergo stunning transformations on their way to adulthood...

## Complete metamorphosis
Butterflies, ants, beetles...

**Egg**
The embryo develops.

**Larva (or caterpillar)**
The larva looks nothing like the adult. Its habitat and food can also be different.

## Incomplete metamorphosis
Grasshoppers, dragonflies, bugs...

*Carolina grasshopper*

**Egg**
The embryo develops.

**Larva**
There is no nymphal stage. When it leaves the egg, the larva already looks like the adult, but without wings. It lives its life very much like the adult, and it grows by molting several times.

**Adult (or imago)**
The larva has finished its transformation and its wings have developed.

Painted lady

### Nymph (or chrysalis)
When the time is right, the larva builds itself a cocoon. Its organs become liquified and give way to a new organism.

### Adult (or imago)
The insect emerges from the nymph all grown up!

## Viviparous incomplete metamorphosis
Aphids, cockroaches...

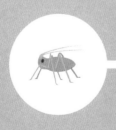

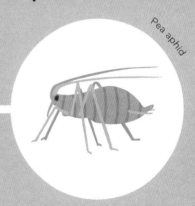

Pea aphid

### Larva
There is no egg—the larva is already formed at birth. It lives its life much like the adult. To grow, it molts several times.

### Adult (or imago)
The larva has finished its metamorphosis!

# Reproduction

Insects reproduce in various ways, and sometimes a mate isn't even required...

## As a couple...

Generally, reproduction in the animal kingdom is **sexual**: a male and a female have intercourse.

## ...or as a solo act!

The females of some insect species, such as stick insects, cockroaches, and aphids (or other animals, including some reptiles and fish) are able to reproduce all alone through a process known as **parthenogenesis**. Their reproductive systems can produce eggs without fertilization.

### Acrobats

Back-to-back, like the small milkweed bug, or in a heart-shaped formation, like damselflies—insects can be true acrobats when it comes to mating!

### How the aphid does it

In the fall, aphids reproduce in a sexual manner, involving a male and a female. The insects overwinter as eggs. In spring and summer, the "viviparous" (live-birthing) female reproduces all alone, producing only females. These new females then give birth to males and females, who go on to repeat the cycle.

# Egg laying

Long or round, white or colorful, single or one of a clutch... insect eggs are extremely diverse.

Yellow and oval-shaped, seven-spotted ladybug eggs are laid in small piles on leaves.

Painted lady eggs are tiny and green, and are laid one by one on leaves.

Praying mantis eggs develop in a sort of protective shell called an "ootheca."

The female Carolina grasshopper extends her abdomen into the ground to lay her elongated eggs.

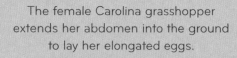

Suspended at the end of a filament, golden-eyed lacewing eggs literally hang by a thread.

The walking flower mantis is well concealed among the flowers.

With its menacing proboscis, the dark-edged bee-fly passes itself off as a bumble bee; yet it's just a harmless fly.

Thorn bugs go about undetected in rosebushes.

The margined calligrapher is yellow and black to trick predators into thinking it's a wasp.

The lappet moth resembles an oak leaf, and this is reflected in their scientific names: *Gastropacha quercifolia* (lappet moth) and *Quercus* (oak).

The buff-tip moth looks a lot like a twig.

# Camouflage and mimicry

Insects have developed a number of strategies for going about their business unnoticed. Some take on an appearance that resembles their environment, while others have the ability to imitate other species or change colors and shapes.

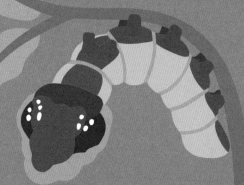

In its larval form, the moth species *Hemeroplanes triptolemus* is aptly known as the snake caterpillar.

The male common cockchafer fans out its antennae to pick up the "pheromones" (chemical compounds) of the female.

# Communication and seduction

Insects have many different ways of communicating, which allow them to share vital information and entice partners during mating season.

En garde! Male eastern dobsonflies jaw joust to defend their territory.

The male spring field cricket rubs his elytra together to emit a high-pitched chirp that is irresistible to females.

Ants transmit information to one another by rubbing antennae.

A male scorpion fly shows affection to his ladylove by offering her a fly.

The firefly emits a luminescent signal from the end of its abdomen to attract a mate.

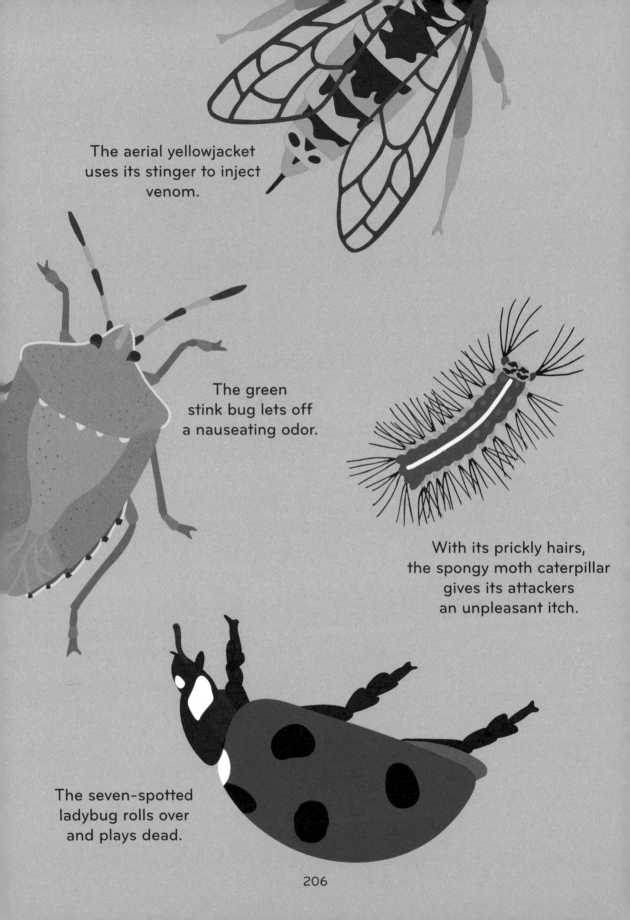

The aerial yellowjacket uses its stinger to inject venom.

The green stink bug lets off a nauseating odor.

With its prickly hairs, the spongy moth caterpillar gives its attackers an unpleasant itch.

The seven-spotted ladybug rolls over and plays dead.

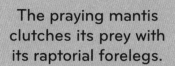

The praying mantis clutches its prey with its raptorial forelegs.

# Defense

They might be small and appear helpless, but insects have no shortage of self-defense techniques when danger comes calling.

The odorous house ant spits formic acid.

The fork-tailed bush katydid will shed a limb to free itself from predators.

# Predators

Many, many living creatures feed on insects: birds, lizards, hedgehogs, bats, frogs, and anteaters, as well as moles, spiders, bears, badgers, humans, carnivorous insects, and even carnivorous plants! If insects disappeared, the entire food chain would be threatened.

# Miraculous migrators

## Why do insects migrate?
Whereas birds migrate to have enough to eat during winter, insects do so because of changes in temperature. Insects are cold-blooded animals, so their temperature depends on their environment. That's why they migrate when the weather is too cold or too hot, depending on the region!

## How do they find their way?
Insect migration remains a largely unstudied and poorly understood phenomenon. However, according to one theory, insects may be equipped with internal clocks and compasses that give them a strong sense of time and space. Other theories hold that insects are attuned to the Earth's magnetic field, the position of the sun, or the direction of the wind.

## Short- and long-distance travelers
While some insects travel dozens or even hundreds of miles to get to a neighboring region, others embark on even more impressive migrations, braving winds, crossing seas and mountains, or even changing hemispheres! Some soar to heights of more than a kilometer (3,281 feet) in the sky and cover distances of more than 4,000 kilometers (2,500 miles).

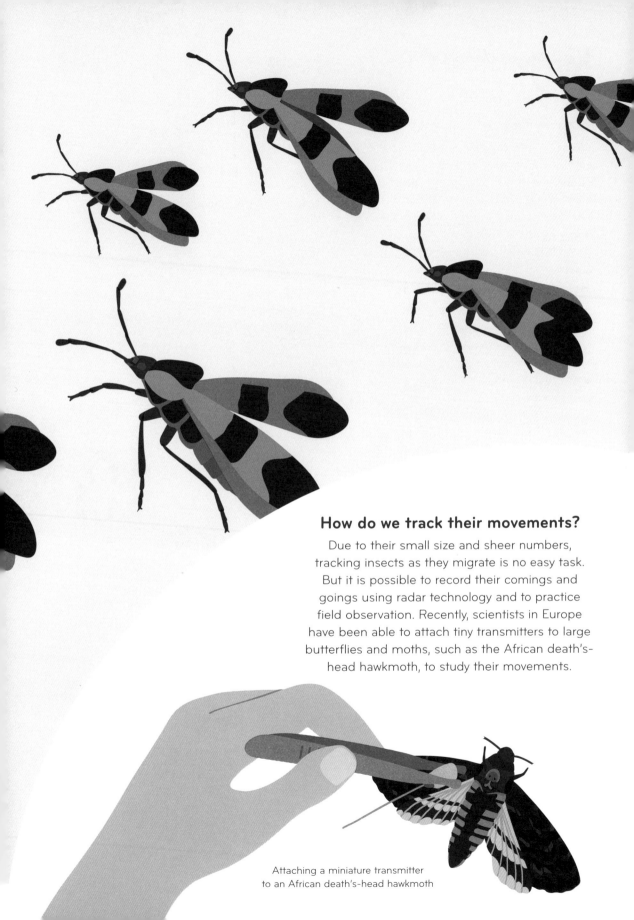

### How do we track their movements?

Due to their small size and sheer numbers, tracking insects as they migrate is no easy task. But it is possible to record their comings and goings using radar technology and to practice field observation. Recently, scientists in Europe have been able to attach tiny transmitters to large butterflies and moths, such as the African death's-head hawkmoth, to study their movements.

Attaching a miniature transmitter to an African death's-head hawkmoth

# Who goes where?

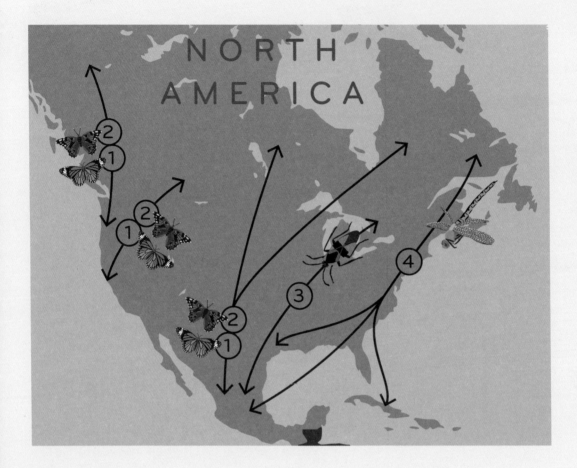

### ① Monarch
### ② Painted lady

Butterflies that spend the summer in eastern North America migrate to Mexico for the winter. Those that spend the summer in the American northwest head down to the California coast. It takes four generations of monarchs to complete the journey, and six generations of painted ladies.

### ③ Large milkweed bug

Bugs that summer in southeastern Canada overwinter in Mexico—a trek that plays out over four generations.

### ④ Green darner

Those that spend the summer in eastern North America overwinter between the southern United States and Mexico, and sometimes the Caribbean. It takes three generations to complete the trip.

# A cross-generational journey

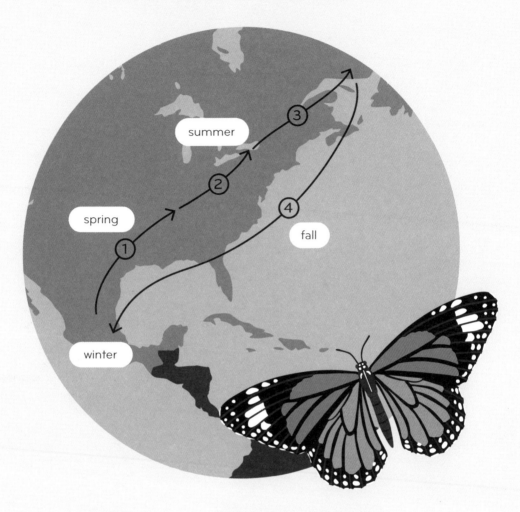

As insects have short lives, completing a long journey often requires a multi-generational effort. Each step of the way, adults lay eggs, eggs produce larvae, and the larvae become nymphs. The winged journey resumes when the nymphs reach adulthood.

**How it works for monarchs**

① A first generation of adults leaves Mexico at the end of winter, stopping over in the southern United States. A new generation is born.

② The second generation of adults flies up the East Coast of the United States, completing another lifecycle in the process.

③ The third generation of adults reaches Canada, where yet another lifecycle comes to an end.

④ The fourth generation heads back to Mexico for the winter.

# Why are insects disappearing?

40% of the world's insect species are threatened with extinction. Here are the main reasons for this alarming trend.

### Construction
The natural habitats of many insect species are being lost to road construction and housing developments. Forests are also disappearing because of the logging industry.

The golden ground beetle is losing its habitat, day by day.

Pollinators—like bumble bees and other types of bees, flies, and butterflies—are becoming scarcer as their food sources continue to disappear.

### Fertilizers and pesticides
Chemical fertilizers and insecticides threaten the existence of numerous species that make their homes in crop fields and grasslands. Because of these products, plant diversity is dwindling and, as a result, so is insect diversity.

Water sources are drying up as the planet gets hotter, which is affecting the chances of survival for the four-spotted skimmer.

## Climate change

Temperatures naturally go up and down with the seasons, telling insects when it's time to migrate or reproduce, among other things. But temperature changes are becoming increasingly erratic due to climate change, which is primarily caused by greenhouse gas emissions from vehicles and factories. Springs are becoming warmer and drier, and winters are getting shorter and shorter, disrupting the natural cycles of insects!

## Light pollution

Streetlights, store windows, and commercial signs are lit up at night. Nocturnal insects are drawn to the light and often either die of exhaustion from flying around the light source or become easy prey for birds and bats.

The spindle ermine moth loses its sense of time because of artificial light.

# What insects do for the planet

They pollinate flowers and help plants reproduce. They play a major role in the production of fruit, vegetables, coffee, tea, chocolate, and even cotton.

They break down plant and animal matter, recycle and oxygenate the soil, and remove excrement.

They are food for lots of different animals.

# What insects do for us

We use the wax and honey made by bees and the silk from silkworms.

They rid our vegetable gardens of aphids.

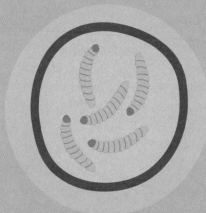

They are a common food for people in Asia, Africa, South America, and, increasingly, in North America.

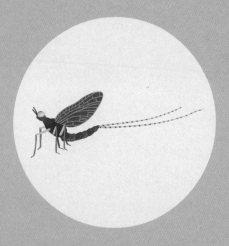

Their presence or absence tells us a lot about the health of our environment.

# Let's do our part

Here are some easy things you can do to help protect insects:

Observe them without disturbing them.

Add an insect hotel to your garden.

Let a section of your garden revert to its wild and native state, and mix up the flowers on your balcony or deck.

Avoid using pesticides and fertilizers.

Ride a bike to help
fight climate change.

Avoid unnecessary
lighting at night.

Cut down on food waste
and purchase organic produce
and products whenever possible.

Take part in environmental
action and make sure
your friends and family
understand how important
it is to protect insects.

For more educational tools and a teacher's guide, visit hello.helvetiq.com/insectorama

## Color in
four different insects

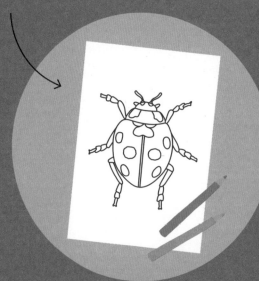

## Test
your knowledge with a memory card game

## Build
your very own insect hotel

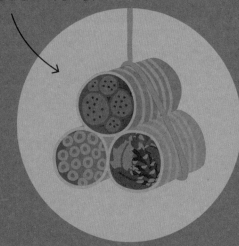

# About the author

My name is Lisa Voisard and I was born in Lausanne, Switzerland, in 1992. As a graphic artist, illustrator, and musician, I have always had a need to create, and through my creations I have found countless ways to express myself!

Creating *Insectorama* was a fun opportunity for me to get an up-close view of nature and dive deep into the secret world of insects. Every living being, no matter how small, has its own life to lead, its own way of doing things, and its own behaviors—a delicate balance that must be respected.

The more we understand our environment, the better we can care for it. So, let's give nature some attention and get to know her a little better!

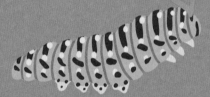
Black swallowtail caterpillar

# Thank you...

To the entire Helvetiq team, for your faith in me and for seeing me through the creation and promotion of *Insectorama*. To Aude Pidoux for managing the project, to Jeffrey K. Butt for his translation skills, and to Angela Wade and Sonia Curtis for their thorough review.

To Gaël Pétremand and Mathilde Gaudreau, both amazing entomologists, who always made themselves available and were kind, jovial, and helpful. And to three more entomologists—Christian Monnerat, Jean-Luc Gattolliat, and Yannick Chittaro—for their help as well.

To Malik Beytrison for his invaluable support. Some of the discoveries made are not mine, but ours.

To my parents and my sister who are always there to cheer me on through my adventures and life choices.

And to all the insects... for teaching me that, sometimes, smallness is greatness!

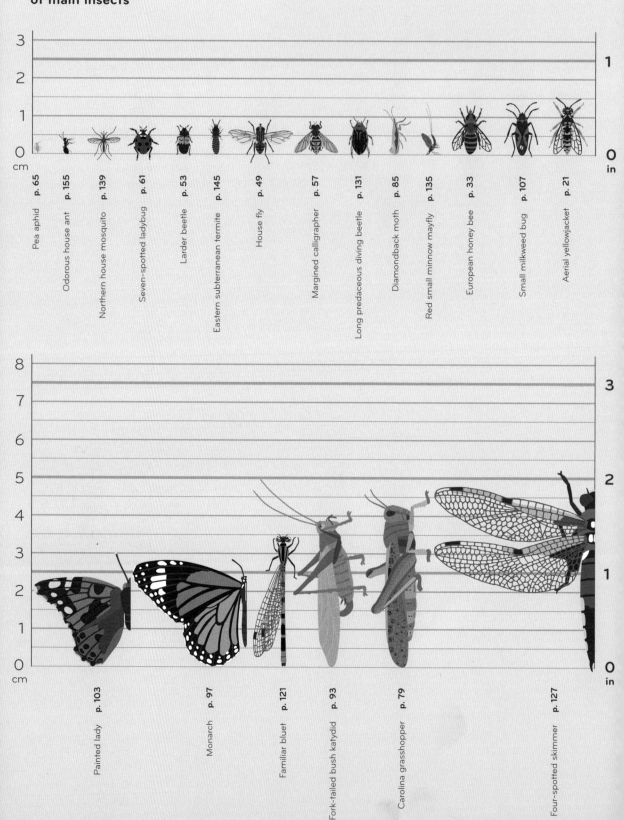

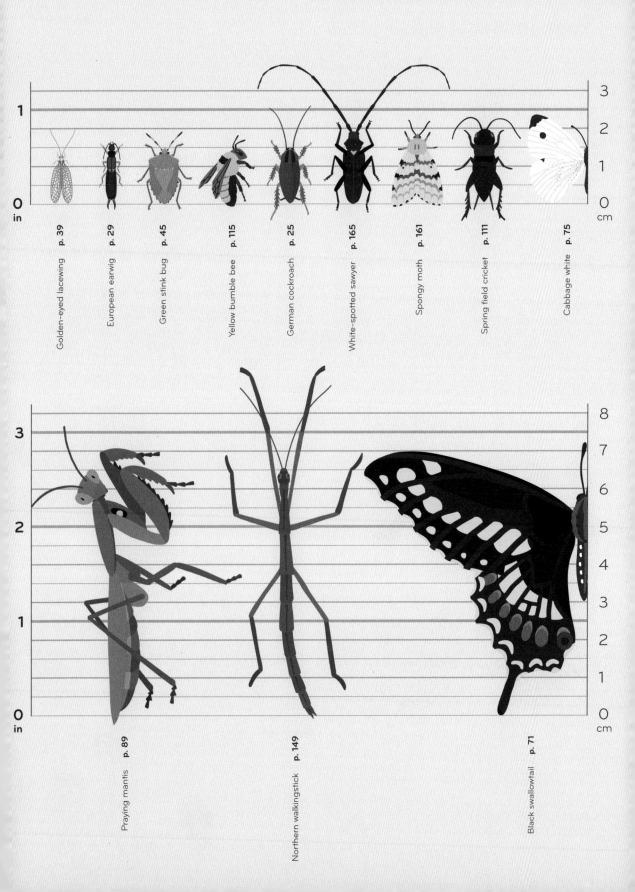

# Alphabetical index

## A
Aerial yellowjacket  21  34  58  206
African death's-head hawkmoth  211
American cockroach  26

## B
Black carpenter ant  156
Black swallowtail  71
Blue bottle fly  50
Brown-belted bumble bee  116
Buff-tip moth  203

## C
Cabbage white  75
Canadian tiger swallowtail  72
Carolina grasshopper  9  79  94  198  201
Chinese mantis  90
Common clothes moth  86
Common cockchafer  204
Common crane fly  140
Common sulphur  76

## D
Dark-edged bee-fly  202
Diamondback moth  85

## E
Eastern dobsonfly  204
Eastern subterranean termite  145
European earwig  29
European elm bark beetle  192
European honey bee  22  33
Eyed click beetle  166

## F
Familiar bluet  121
Firefly  177  205
Forest giant owl  172
Fork-tailed bush katydid  93
Four-spotted skimmer  127  215

## G
Gall wasp  193
German cockroach  25
Giant walkingstick  150
Giant water scavenger beetle  132
Golden-eyed lacewing  39
Golden ground beetle  214
Green burgundy stink bug  46
Green darner  174  212
Green stink bug  45  206

## H
Harlequin ladybug  62
*Hemeroplanes triptolemus*  203
Hercules beetle  169
House cricket  112
House fly  49

## L
Lappet moth  203
Larder beetle  53
Large milkweed bug  108  212
Leafcutter  192
Long predaceous diving beetle  131

## M
Margined calligrapher  57  202
Michigan hex burrowing mayfly  136
Migratory locust  80
Monarch  97  212  213
*Myrmeleon immaculatus*  176

## N
Northern house mosquito  139
Northern walkingstick  149

## O
Odorous house ant  146  155

## P
Painted lady  98  103  199  212
Pea aphid  65  199
Pecan weevil  192
Periodical cicada  171
Pharaoh ant  156
Posterior brown lacewing  40
Praying mantis  89  207

## R
Red admiral  104
Red small minnow mayfly  135
Rose aphid  66
Rove beetle  30

## S
Salt marsh moth  162
Scorpion fly  178  205
Seven-spotted ladybug  61  206
Small milkweed bug  107
Snowberry clearwing  179
Spindle ermine moth  215
Spongy moth  161  206
Spotted spreadwing  122
Spring field cricket  111  205

## T
Thorn bugs  202

## V
Varied carpet beetle  54
Viceroy  98

## W
Walking flower mantis  202
Wandering glider  128
Water strider  170
White-spotted sawyer  165  193

## Y
Yellow bumble bee  115